PANAMA CANAL TOWNSITES

This book, originally published by the Panama Canal Museum, is reprinted in honor of the Museum's history and that of the people who lived and worked in the Panama Canal Zone.

LIBRARY PRESS@UF

AN IMPRINT OF UF PRESS AND
GEORGE A. SMATHERS LIBRARIES

Panama Canal Townsites

Panama Canal Museum

Library Press @ UF
Gainesville, Florida

Names: Panama Canal Museum, publisher. | George A. Smathers Libraries, publisher.

Title: Panama Canal townsites / presented by the Panama Canal Museum.

Description: Gainesville, Fla. : Library Press @ UF, [2015] | Title from pdf. | Includes bibliographical references (page 160). | Originally published in 2011 by the Panama Canal Museum. | Library Press @ UF: an imprint of University of Florida Press and George A. Smathers Libraries. | Contents: Construction era towns—Ancon—Balboa, Balboa Heights—Cardenas—Cocoli—Coco Solito—Coco Solo—Cristobal, New Cristobal, Old Cristobal—Diablo, Diablo Heights—Gamboa—Gatun—La Boca—Los Rios/Corozal—Margarita—Paraiso—Pedro Miguel—Rainbow City—Red Tank—Military towns. Albrook Air Force Base; Curundu, Curundu Heights; Farfan Naval Radio Station; Fort Amador, Fort Grant; Fort Clayton; U.S. Naval Air Station at Coco Solo; Fort Davis; Fort Gulick; Fort Kobbe; Howard Air Force Base; Fort Sherman; France Field; Quarry Heights; Rodman Naval Station.

Identifiers: ISBN 9781944455026 | ISBN 1683400089

Subjects: LCSH: Cities and towns—Panama—Canal Zone—History. | Cities and towns—Panama—Maps. | Cities and towns—Panama—Canal Zone—Maps. | Canal Zone—History. | Canal Zone—History—Pictorial works. | Canal Zone—Social life and customs.

Classification: LCC F1563

Library Press @ UF is an imprint of the University of Florida Press.

LIBRARY PRESS @ UF

AN IMPRINT OF UF PRESS AND
GEORGE A. SMATHERS LIBRARIES

University of Florida Press
15 Northwest 15th Street
Gainesville, FL 32611-2079
http://upress.ufl.edu

Series Foreword

The Panama Canal Museum, formerly located in Seminole, FL, closed in 2012 and transferred its collection to the George A. Smathers Libraries, greatly enhancing the University of Florida's holdings on Panama and the Canal. The Museum's mission was to document, interpret, preserve, and articulate the leadership role played by the United States in the history of the Panama Canal, with emphasis on the construction, operation, maintenance, and defense of the Canal and the contributions to its success by people of all nationalities. This mission continues to guide the preservation of the Panama Canal Museum Collection. As such, the Smathers Libraries have reprinted this book, originally published by the Panama Canal Museum, in honor of the Museum's history and that of the people who lived and worked in the Panama Canal Zone.

The Panama Canal Museum Collection (PCMC) in the Department of Special & Area Studies Collections at the George A. Smathers Libraries is the leading research collection in the United States for the study of the American era of the Panama Canal. The Collection documents the U.S. experience in the Panama Canal Zone and Panama, and to a lesser degree, it also preserves historical information about the Canal prior and subsequent to U.S. construction and operation.

The Smathers Libraries preserve and provide access to the historically significant and distinctive materials about Panama and the Canal in order to facilitate knowledge creation and dissemination and to support and advance the commitment of the Smathers Libraries to excellence in education and research and contribute to the University of Florida's standing as a preeminent public research university.

Visit the Panama and the Canal Digital Collection at *http://ufdc.ufl.edu/pcm*.

Judith C. Russell, Dean of University Libraries, University of Florida

Table of Contents

Preface

When the United States took on the building of the Panama Canal in 1904, workers were faced with extremely difficult living conditions. The tropical diseases such as malaria and yellow fever still plagued them just as they had the earlier French effort. The housing stock left behind by the French was dilapidated and inadequate. The French had turned over 2,265 buildings, scattered at 79 different points along the line of the Panama railroad, over a distance of more than 60 miles from Colon on the Atlantic side to Panama and adjacent islands on the Pacific side. But the existence of a great many of these buildings was apparent only from the records. They had been overgrown by vegetation and quickly lost to the jungle. About a hundred sets of beautifully drafted architectural plans left by the French came in handy for locating drains, etc., as the Americans made repairs to existing buildings.

Some workers found insect ridden rooms in adjacent towns while others lived in tents or thatched huts near construction sites. Not wanting to endanger the lives of their families, most men left their wives and children behind. For many of these workers, who had left home for the first time, the loneliness was excruciating. Men, like Madison Fletcher Bradney (p. 9) waited for the mail boat to bring news from home and counted the days until their sweethearts would join them.

The first Canal Zone governor, General Davis, was opposed to the presence of American women on the Isthmus. The men were going to build the canal, he said, not the women.

But soon it was discovered that the married men would not stay for any length of time without their wives and families, and as they were more to be depended upon than the single ones, the authorities decided to overhaul the old French houses and make them as comfortable as possible under the circumstances for the families who wished to come down, and also granted them special steamship rates.
- Mrs. Chas. C. J. Wirz

As Harriet Verner explained, "The men came for money, for adventure, for fame, or in sheer wanderlust—the women came for love . . . ." Love must have been the only thing that kept them there. Despite the renovations to the old French quarters and hospital wards, living conditions were deplorable,

My husband, Dr. Darling, sailed on Feb. 21, 1905, for the Isthmus of Panama, three days after our wedding. Leaving me a forlorn bride in N.Y. City,--there being no available quarters for American women. The following October my husband called me to come. I sailed on the Finance, Oct. 3, 1905. In reaching Colon, I boarded the only train for Panama City,--through a dense jungle of riotous tropical growth. I am sure I felt as much depressed as Columbus felt elated in this new

found land. On my arrival in Panama, my husband and I lunched at the Hotel Central and thence to the secured quarters at Ancon Hospital, which was one decidedly spacious room. It had only accommodated twenty-four patients previously! Furnishings consisted of one iron bed, boxes for dresser and wash-stand; one three-legged chair, and a mildewed mirror hung on the wall. No lights whatsoever! I was fortunate in finding a lantern on the road side one night. With no scruples whatever, I took it home.
- Nannerl L. Lewellyn Darling

Rose Van Hardeveld moved with her family into a renovated French cottage, House No. 1, (see p. 102) which she said "contained only three small rooms and kitchen with veranda around three sides and the only way to get from the house to any other part of the village was to walk on the railroad track, an extremely dangerous thing to do, because of the trains constantly in motion, also the sharp curves in the track just here. Hardly a day passed without someone getting hurt or killed almost at my front door."

When John F. Stevens took over as Chief Engineer in 1905 he took immediate steps to improve living conditions in the Canal Zone, while putting aside the directive of the Isthmian Canal Commission "To Make the Dirt Fly." Under the direction of Chief Sanitation Officer, Dr. William Gorgas, the Canal Zone and immediate vicinity was cleaned up. Before sewers and water mains were installed, people had to collect rainwater in tanks or pay 15 cents for 5 gallons from water wagons vendors. Streams, ditches, and open containers, where disease carrying mosquitoes could breed, were sprayed or removed. The existing housing was retrofitted with screens on the windows and doors and new ones built to Gorgas' specifications. According to Mrs. Wirz, "The Sanitary Department used to fumigate all the houses in Panama very frequently. They would close ev-

ery opening, every crack in and outside with paper so that no fumes could escape and rooms remained hermetically closed for hours. Much grumbling could be heard on the account of the damage the sulphur [sic] did to mirrors and other articles. After a time the yellow fever abated and people were left in peaceful possession of their homes."

It soon became evident that a great number of new buildings would be required for living quarters, both for single and married employees, black and white, wrote P. O. Wright, Jr., architect in The Canal Record, December 11, 1907, "These buildings had to be designed not only to meet the requirements of climate but for material which would stand shipment from the United States with the least chance of being ruined by breakage or exposure to the weather. The character of the building sites and means of transportation of material to the same, the official status of employee, the cheapest methods of construction and an effective method of screening were other matters that had also to be considered."

By the beginning of 1908, work was underway to provide living accommodations for married employees. New

Former French cottage improved under the American administration and assigned as married quarters. *From the Hallen Collection.*

houses under construction at Cristobal and Empire were completed relieving some of the demand for married quarters. The shifting character of the work led to an apparent surplus of bachelor quarters at one or two points along the line. These former bachelor quarters were converted to new houses two, four, and even eight families.

A racially segregated system known as the Gold and Silver Roll was implemented by the American administrators of the Canal Zone in 1904 and became the foundation for Panama Canal Zone society and economy until it was phased out in the 1960's. The Gold and Silver Roll system in the Panama Canal Zone was more than just a pay system designed to maintain a more privileged class of white semi-skilled and skilled workers happy with their stay in Panama.

The Gold Roll enjoyed all of the privileges and amenities that the system had to offer. They enjoyed, of course, much higher pay, better and more spacious housing facilities for families, excellent and well equipped schools for their children, better nutrition, better health care, almost free entertainment and recreational facilities and a generally better quality of life. Their (the Gold Roll) comfort and satisfaction were central factors in most decisions made by the Canal administrators.

Other benefits that became very important "draws" in the recruitment process were sick leave and "home" leave, a privilege that included paid return passage back to their home state for a holiday while their job was preserved for them on the Zone. Although some non-American members of the Gold Roll were entitled to the "privileges" of this special group of people, they were, nevertheless placed at a lower pay scale and denied certain benefits, particularly, sick and home leave.

The Gold Roll, paid in American gold dollars, reflecting a much higher pay scale than in the U.S., at first was comprised of chiefly white American employees brought in from the United States mainland. The Silver Roll or "local-rate" employees were mostly West Indians brought to the Canal Zone to work as laborers and in the low paying jobs throughout the Zone. They lived in separate housing areas that would become small cities in their own right, like Rainbow City and Red Tank.

The following information about quarters for laborers was excerpted from articles in the the *Panama Canal Record* of 1907-1908.

QUARTERS FOR LABORERS AND "SILVER" EMPLOYEES

In providing buildings for the use of laborers or "silver" employees engaged in the construction of the Canal, special consideration has been given to their design with a view of meeting the requirements of the different nationalities and their habits of living. This applies both to the married and single employees.

Special consideration has also been given to the sanitary features of these buildings as well as to the location of them, also to the methods of protecting the mosquito screening, the doors and the windows, against careless and especially against willful injury. The following buildings are provided, where the conditions will warrant their erection, in each of the many encampments scattered along the line of the work from Colon to Panama: Bachelor quarters or barracks; married quarters; bath houses for colored and for European employees; laborers' kitchens and Gallegos' mess halls; automatically flushed range closets for men and for women; dry and wash rooms; and sick camps. The sick camps are for the immediate care of sick employees until their removal to the hospital.

In addition to the above buildings, which are intended principally for the encampments of laborers employed in the actual work of digging, there are certain other types of buildings erected for the employees connected with hotels, hospitals, etc. These buildings are constructed in the cheapest possible manner consistent with their uses and are all provided with sanitary plumbing. Among

the types may be cited the following: Ward maids' quarters for the hospitals; orderlies' quarters for the hospitals; quarters for the hotel servants, both one and two-story buildings; and quarters for the help of commissaries.

FIRST SCHOOLS

One of the first teachers in the Canal Zone, Mrs. W.E. Maxon, set about the challenging work setting up a school in Las Cascadas in April 1906. Upon arrival on her first day, she discovered,

"a bare room 24x36, not even a blackboard installed but about 26 pupils of all sizes and colors awaiting us. Some accompanied by their parents and some by grandparents. Each one had some advice to give concerning discipline of

their children and even furnished the straps to lash them with. I told each child after enrollment to bring a bottle of water and a box to sit on when they returned in the afternoon, as there were no desks or chairs and for three weeks most of them sat on the floor, and I used a large slate for a blackboard. There were only three American families in Las Cascadas at that time and one of them kindly loaned me a chair and table for a desk."

CLUBS

As quality schools were built to educate future generations of Zonians, churches were opened to sustain their spirit. Clubhouses became the social centers for Canal employees and their families. After his visit in the spring of 1907, Secretary of War, William Taft, sent Helen Varick Boswell from the General Federation of Women's Clubs in New York to the Canal Zone to organize clubs to alleviate "the discontent and loneliness of some classes of the women." The Canal Zone Federation of Women's Clubs held meetings to inspire its members who took on such projects as a circulating library and sending Christmas presents to the leper colony at Palo Seco and to prisoners. The Canal Zone Federation was disbanded in 1913.

Another club that prospered in the Canal Zone was the Y.M.C.A., which opened its doors to employees' families. The men enjoyed a reading room, a game room, and writing desks. Women could bowl, exercise and join a "Friendly Circle." There was scouting and even a movie from time to time in those early years. Residents also participated in ball games, swimming, boating, and dinner parties.

NEWER HOUSING

Over the years townsites were developed on both the Pacific and Atlantic sides of the Isthmus to meet the needs of Canal workers and their families. From the construction days forward, new communities were developed and/or improved.

Quarters for Silver Roll employees in the early construction days are scattered in front of the newly-built Tivoli Hotel. They were eventually renovated or replaced by screened-in barracks and homes, circa 1912. *From the Hallen Collection.*

In 1952 a new round of building and replacing aging living quarters was initiated throughout the Canal Zone. It was an exciting time for families who took the opportunity to "house shop." As many families employed the services of West Indian maids, some of the new houses also had a built-in "maid's quarters."

The assignment of housing in the Canal Zone was based less on rank than on years of service, family size, job assignments. Employees could request new quarters on a weekly basis but it might take weeks, even months, for a place to become available. Applicants would check the bulletin boards weekly to see if their turn had come up and they could move.

Some workers were required to live in assigned housing in the towns where they were employed. For instance, those working in the Dredging Division lived in Gamboa, where the headquarters was located. Sometimes large families were assigned two units in order to accommodate all their children. Unmarried men and women were assigned to bachelor quarters, which started out as small "shot-gun" style units, where they had to pass through the bedroom to get to the kitchen. Enventually, they could "graduate" into a more conventionally designed one-, or even, two-bedroom unit.

What had started out as a cesspool of disease and loneliness, emerged as a little piece of paradise for its Canal Zone residents. This book tells some of the stories of the various townsites scattered along the fifty miles of the Panama Canal Zone between the Atlantic and the Pacific Oceans. It also shares the fond memories of a few of its residents whose hometowns have changed since the Panama Canal was turned over to Panama on December 31, 1999, and the Canal Zone as they knew it was no more.

Several years ago, Robert Will suggested that the museum consider displaying Canal Zone townsite maps at the Panama Canal Society's Annual Reunion. Robert's idea was that former residents of the Canal Zone could write their names on the houses to show where they lived in each town. The maps have proven to be a highly popular and interactive display over the past few years.

Noting the popularity of the display, Jerry and Sharen Halsall suggested to museum president Joe Wood during the 2010 Annual Reunion that the museum take it one step further and publish a townsite book. After a year in the making, we are proud to present our latest publication – we hope you enjoy it. We are grateful to Robert, Jerry and Sharen for sharing their wonderful ideas.

We would like to recognize the special contributions of photographs by Carl Berg, Vicki Hutchison Boukalis, Dick Cunningham, Bill Fall, Bob Karrer, Ed Ohman, Joan Ohman, Bob Panzer, Elaine Stevenson and the Ernest "Red" Hallen Collection.

We would also like to thank our book committee members, Cheryl Russell, Chair; Elizabeth Neily, layout and design; Kathy Egolf, proofreader; Tom Costello and Dick Cunningham, scanning; and Isabel Wood Egan, Ann Wood, Judy McCullough, Naomi Litvin and Marian Blair for their hours of researching and typing stories.

Lastly we want to thank all the Zonians who contributed their personal stories, photographs and financial support which have turned this book into a wonderful treasure.

New Types of Canal Housing Attracts "Window Shopping" House Hunters

Panama Canal Review, September 5, 1952, p.16.

House-hunting is becoming a common diversion among Canal employees now that 1952 housing projects are turning into walls and roofs and windows that potential occupants can see in their real settings. Many new types of houses have not been seen before in Canal communities.

The new patio house (photo 1) is being built both in Margarita and in Ancon. It is a two-bedroom house with the main feature being the central covered patio and elevated roof which lets light and air into the dining-living room. A duplex version of the patio house can be seen without a roof in the second photo. The third photo shows a new type cottage which is a comparatively small three-bedroom house with two baths and a maid's room and bath. It has an L-shaped living room with dinette space. The only four bedroom house in this year's building program is a breezeway house (Type 329) shown in the bottom picture.

Left, Patio type house in Ancon nearing completion.

Below Left, General view of new Ancon development.

Below Right, New type three-bedroom cottages in Ancon and Margarita.

In 1910, my family, the Nevilles, resided in this house in the town of Bohio. After Gatun Lake filled up, the house disappeared under forty feet of water! *Courtesy of Malcolm Stone.*

4

Construction Era Towns

Sponsors

In Memory Of

Gerald (Roosevelt Medal & Three Bars Holder) & Mabelle Bliss
(by Mickey Walker Fitzgerald) –
Allen Stepp Boyd (Roosevelt Medal & Two Bars Holder)
(by Barbara Bartholomew Krueger)
M. F. Bradney (Roosevelt Medal & Three Bars Holder) & Stella Vaughen Bradney
(by their Egolf & Wedwaldt grandchildren & great grandchildren)
Charles C. (Roosevelt Medal & Three Bars Holder) & Alice Pierce Clement
(by their grandchildren)
Charles D. Hummer (Roosevelt Medal & Four Bars Holder)
(by Charles W. "Chuck" Hummer)
Benjamin Davis (Roosevelt Medal & One Bar Holder) & Mary Regina (Marie) McConaghy
(by their granddaughter Kathleen McConaghy Campbell)
Walter D. Peterson (Roosevelt Medal & Three Bars Holder)
(by Cheryl Peterson Russell)
Charles E. Thomas (Roosevelt Medal & Two Bars Holder)
(by J. E. Dorn Thomas)

Canal history . . .

Chagres River Dam creates Gatun Lake, floods old townsites

The Panama Canal Spillway, June 14, 1996, vol. XXXIV, no. 12, p. 2.

The following article, taken from The Panama Canal Record of December 6, 1911, tells about the settlements that would be covered by water with the forming of Gatun Lake following the construction of Gatun Dam. The article provides an interesting look at the history of the Panama Canal area and into life on the Isthmus during Canal construction days. Because of its length, the article will be published in five parts.

Part 1

Early village
Thatched roofs are interspersed with new metal ones and cane houses in the village of Cruces in this 1912 photo. Cruces was the last landing for canoes traveling up the Chagres River from the Atlantic side, after which travelers had to go overland to reach the Pacific.

The villages between Gatun and Matachin will be covered by the water of Gatun Lake. They have never been important in the sense of size, or as the center of any peculiar type of life. In fact they are little more than jungle hamlets, yet they have a distinct place in American history, because they were known to European civilization many years before Jamestown (the first settlement in North America, founded in 1607) was settled or Massachusetts Bay (joint stock company set up by royal charter in 1629) was an English colony.

In *The Panama Canal Record* of November 29, there was republished a letter (from Gen. George W. Davis, Isthmian Canal Commission member) in which attention was called to the fact that the names of some of these villages appeared on the map published with Esquemeling's (Alexander Esquemeling) narrative of the Buccaneers in 1678. Most of them antedate that time, for they were not named by the English who plundered with Morgan (Sir Henry Morgan, known for destroying Panama City in 1671), but are spoken of in Esquemeling's book as places already known, and invariably they bear Spanish names. It is probable that most of them date from the early days of navigation on the Chagres River, when it was one of the most used routes for commerce across the Isthmus. Among these are Ahorca Lagarto, Barbacoas, Caimito, Matachin, Bailamonos, Santa Cruz, Cruz de Juan Gallego and Cruces (Venta Cruz).

As early as 1530 Spanish ships sailed down the coast from Nombre de Dios and entered the Chagres, whence their goods were transferred to canoes and taken up the river as far as Cruces, a distance of 36 miles from the river mouth, near the point where Culebra Cut begins. From Cruces they were taken overland to Panama. At times of high water, when the stream could be navigated readily by shallow boats, this was the easiest route across the Isthmus, although the trails from Nombre de Dios and, after 1586, from Porto Bello were kept open and were much used by pack trains. The harbor at the mouth is not so safe as those of Nombre de Dios and Porto Bello, and yet that the trade by this route was not inconsiderable is attested by the fact that the entrance to the Chagres was guarded by a fort (San Lorenzo). The river hamlets were of the type of the settlements that grew up along the highways during the days of travel by coach and saddle, and their people probably subsisted as much by the trade they drove with travelers as by the products of their own fields. Yet Esquemeling speaks of cultivated fields, so there was undoubtedly some farming along with the travel trade.

The river trade became steadily less after the reign of Philip II, because Spain's monopoly was gone, and the all-water route to Peru by the Strait of Magellan was found less dangerous. But this was because the trade itself was less, for the Chagres route continued in use up to the time of the completion of the Panama railroad in 1855. Since then the villages in the lake region have been "way stations," with two brief periods of prosperity, one when the French were working nearby them, and the other when the Americans were carrying on their operations.

The region in which these lake settlements are situated will probably not be under water before August, 1912, but the railroad track will be torn up in February, and therefore the native hamlets and American canal settlements are being moved, the houses torn down to be erected again elsewhere, or in the case of shacks merely abandoned in the jungle. It is difficult to persuade some of the inhabitants that the inundation will ever take place. One old bush settler, after receiving repeated warnings heedlessly, ventured it as his opinion that the Lord had promised never again to flood the earth. Such people as they will be assisted in their moving, because the present hamlets will be isolated when the railroad is torn up and in case of a sudden rise in the river, with the backing up of water after the Gatun spillway dam is raised, it would be difficult to rescue them.

In this blotting out of the river hamlets and of one of the world's historic trade routes, nothing of value will disappear; only a few shabby hamlets, and a hundred or more isolated huts in the jungle; while the river route will give way to the Canal, and the railroad to a straighter and better line outside the lake area above all danger of flood.

Gatun landing
The old village of Gatun bordered the Chagres River, as seen in this historical photo taken in 1907. Many such villages were subsequently flooded when, during construction of the Panama Canal, Gatun Dam blocked the Chagres and formed Gatun Lake.

Waiting for the train at Matachin Railway Station.

Postcard courtesy of Robert Karrer.

Gatun Lake claims buildings of many sizes and shapes

The Panama Canal Spillway, June 28,1996, vol. XXXIV, no. 13, p. 5.

Part 2

Housing varieties
Nestled at the jungle's edge, various types of housing can be observed in this 1912 photograph of Chagres, a village located at the mouth of the Chagres River. The community included open huts with thatched palm roofs; a sturdy-looking, white-trimmed house with a tin roof and homes built with scrap lumber and thatched palm roofs. At the photo's far left stands a village outhouse.

(to be continued . . .)

This section describes the types of buildings found in the villages that would be flooded in the making of Gatun Lake. From the Canal Record, December 6, 1911.

In the hamlets and the jungle there are three distinct types of buildings, in addition to the quarters for Canal employees. Of these the most picturesque and primitive is the open hut in the jungle, which consists of a palm thatch raised about eight feet above the ground on bamboo poles. Here a bush family has its incongruous being, for this jungle home is often within sight of the railroad trains, and within it one sees plantain being fried in a modern kettle over a modern brazier, while the drinking water is dipped with a gourd from a square, gallon capacity, oil can. A little more advanced type of dwelling is the pretty hut made of closely set bamboo sticks, sometimes plastered with mud, and with the broad overhanging thatched roof, in which lizards and bugs rustle about day and night. There are none of the more substantial native huts, found in some of the villages in the interior of Panama, built of clay blocks and covered with overhanging pantile roofs. The third type of house, although more modern, can scarcely be considered an advance on the bamboo hut. It is built of lumber and covered with corrugated iron roof. Old residents of the Isthmus say that this type is due to the easy pilfering of lumber and roofing iron, left in storehouses and on isolated buildings by the French canal builders, and that it was unknown before 1885. Usually these buildings have been arrested in dissolution by patches of soap boxes or tin flattened out from old cans, which gives them a motley look. The village stores are little better than this latter type of dwelling. Here and there one sees in the settlement of such nondescript houses, the trim little cottages built by the French and more recently used by the American; and the more airy and well-screened quarters of the American canal period. These, however, are late additions. The original villages were jungle settlements existing because of the isthmian transit.

Family members rest in the doorway of their humble home, a mud hut that will be claimed by water with the creation of Gatun Lake.

Memories

Excerpts from Diaries of a Construction Worker

1907

May 20: Took one suitcase to Ancon with me. Move tomorrow. This is my last night as a resident of Paraiso.

May 21: Moved to Ancon today. No pillow on bed. No nails or hooks to hang anything on.

May 22: Eaten alive by bed bugs last night although the mattress was a new one.

June 10: 11 months ago today since I landed on the Isthmus.

1908

January 1: At Panama all day. Went to the Field Sports forenoon. A rather dull day. Would have liked to spend it with the Girl in Ohio.

April 17: Good Friday and a holiday. – and I went to Pedro M. on early train, then walked to Arraijan, just outside the Zone line. A native town exclusively thatched houses of bamboo.

May 12: Am no. 17 on list of applicants for married quarters. Don't know whether they'll reach me by Nov. or not.

May 22: 2 white men killed, one injured, several negroes killed and hurt by explosion at San Pablo this P.M. 26 tons dynamite touched off by lightning.

August 18: I have been assigned quarters but do I need them? Two boats with no word from S.

Sept. 1: Assignment of family quarters at Culebra—non-housekeeping--came today.

Oct. 6: Left Colon at 2 P.M. A fine night and a fine boat. Coast in sight all afternoon.

Oct. 31: Stella and I were married at 6 P.M. by Rev. Smith. Got license this forenoon.

Nov. 22: Worked till noon. Stella went to Sunday School. We walked about the town some in the afternoon.

Nov. 29: Worked in forenoon. Took Stella for a "tramp" to the Rio Grande reservoir and round hill. Home in afternoon. We were at church tonight.

- Madison Fletcher Bradney, husband of Stella Vaughen Bradney and father of Anna Bradney Wedwaldt and Mary Bradney Egolf. Excerpts courtesy of grandaughter Katherine Egolf.

Roosevelt Medal #1604 and 3 bars awarded to Madison Fletcher Bradney for the years he served the Isthmian Canal Commission between 1906 and 1914.

Below, the Union Church Sunday School at Culebra, Canal Zone, December 31, 1911. *Courtesy of Bradney grandson Albert W. Wedwaldt.*

Front and back view of the ICC employee badge # 655 belonging to Madison Fletcher Bradney.

Union Sunday School, Culebra, Canal Zone, December 31, 1911.

Maybe the reason there are not many men in the Union Church photos is they were like my grandfather, R.B. Potter. He attended only the church suppers because the potluck dinners were So Great! *Comment courtesy of Dick Cunningham.*

Gatun Dam buries old Gatun village on Chagres River flats

The Panama Canal Spillway, July 12, 1996, vol. XXXIV, no. 14, p. 5.

Part 3

Chagres River closes
On August 10, 1907, a curtain of dirt spills from a railroad car, positioned on a trestle street townsite of Gatun, adding to an already growing earthen mound that will close the river at the Dam and Gatun Locks.

Here, the author begins to describe the individual villages that would be destroyed with the formation of Gatun Lake, beginning with the village of Gatun.

The old village of Gatun, which lay on the river flats below the present town was abandoned in 1908, and the site is now covered by 80 feet of rock and earth under Gatun Dam. At the time it was abandoned, the village contained a church, priest's house, school, a dozen small shops, and ninety or more small houses of all descriptions, from the bamboo hut with palm thatch to the typical sheet iron roof shanty. Most of the buildings were moved to the new townsite, now known as New Gatun. The railroad line also ran through the dam site and as soon as the present line into Gatun was opened, this likewise was abandoned, and the station building was razed. By the middle of 1909 the last vestiges of the old village had disappeared before the encroaching work on the dam.

The antiquity of the place is uncertain, because none of its buildings were of masonry. In his narrative of the pirate Morgan's march to Panama in August, 1670, Esquemeling says "The first day they sailed only six leagues, and came to a place called De Los Bracos. Here a party of his men went ashore, only to sleep and stretch their limbs, being almost crippled with lying too much crowded in the boats. Having rested awhile, they went abroad to seek victuals in the neighboring plantation; but they could find none, the Spaniards being fled, and carrying with them all they had."

The location on the river corresponds to that of Gatun, for six Spanish leagues equal about nine miles, and even if the situation of De Los Bracos is not identical with old Gatun the narrative indicates that the region thereabouts was somewhat settled. It is also known that the Spaniards had erected a fort on a hill 120 feet above the river, overlooking the town, which

was probably one of the outposts they had established at various points along the isthmian trade routes. Evidences of the old fort are found today, and the site is shown on the original land map made for the Panama railroad in 1855. At that time the village had about one hundred buildings of all kinds. Writing of it in 1861 Otis (F.N. Otis, in his "Handbook of the Panama Railroad") says it was a village composed of forty or fifty huts of cane and palm. In the early days of the California immigration it was the first stopping place in the canoe journey up the Chagres, where "bongo-loads of California travelers used to stop for refreshments on their way up the river, and where eggs sold four for a dollar; and the rent for a hammock was two dollars a night."

In 1881 the French chose Gatun as the site for one of the canal residencies, erected machine shops there and built a number of quarters for laborers, calling the new section, "Cite de Lesseps."

Old Gatun depot
Railroad passengers make themselves comfortable any way possible, sitting on rail carts, sitting with legs dangling over the edge of the depot porch, even by sitting on the railroad tracks themselves, in this 1906 photo of the train station located at the original townsite of Gatun.

This continued as a center of the work of excavation until 1888 when all operations ceased, not to be resumed here until 1904.

When the Americans arrived in 1904, Gatun was the center of a comparatively large river trade. Bananas and other produce from the Gatun, Trinidad and Chagres rivers were brought there for transhipment by rail, and for sale. Once a week a shipment of from seven to nine carloads of bananas was made, and on the shipping day, as many as a hundred canoes would tie up at Gatun.

Waters of Gatun Lake cover Isthmian townsites

The Panama Canal Spillway, July 3, 1985, p. 3.

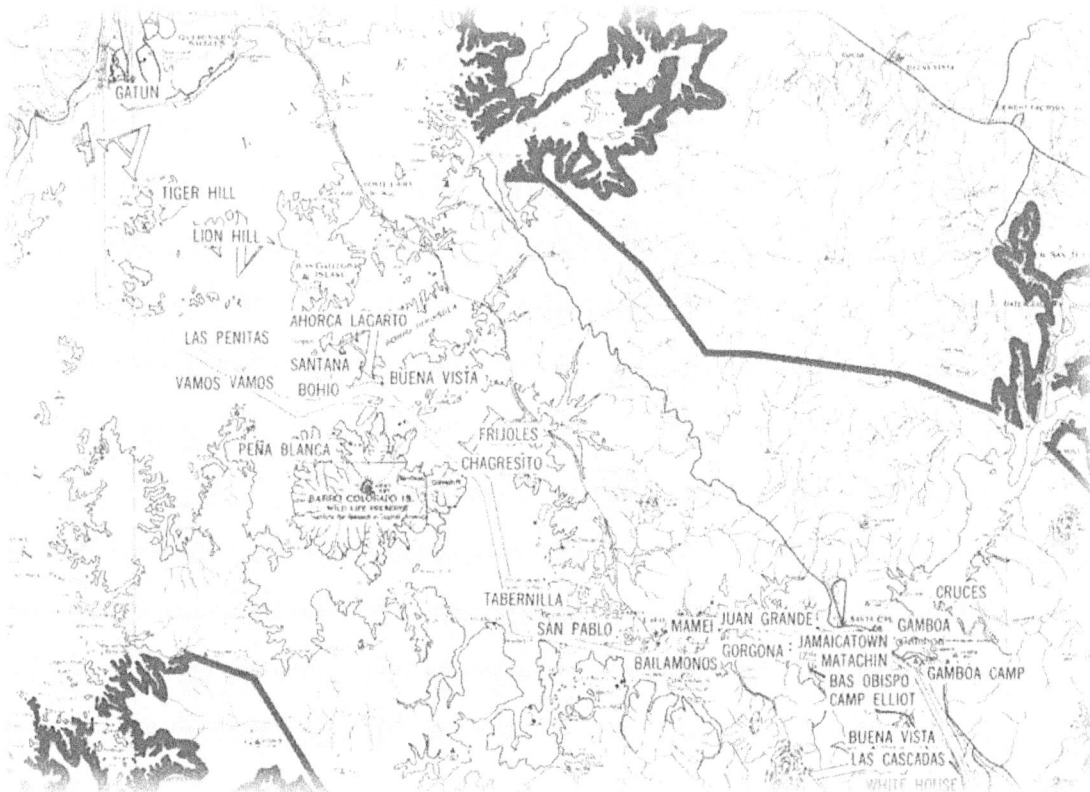

Chagres River covers site of first dam under French construction

The Panama Canal Spillway, July 26, 1996, vol. XXXIV, no. 15, p. 5.

Part 4

Bohio, 1911
In 1911, just up the river from Gatun, the townsite of Bohio sprawls out along the railroad tracks, even including two-story frame homes like the one in the distance. With trains coming and going regularly, the town appeared to flourish, unlike other less-developed villages in the area.

Proceeding on from Gatun, the author continues to describe the significant aspects of the villages to be covered by water during the formation of Gatun Lake.

The next settlement of any importance up the river from Gatun is Bohio. Between these two villages are three hamlets, Lion Hill, Tiger Hill, and Ahorca Lagarto, none of them numbering over half a dozen huts and without any apparent reason for existing except that some bush negroes or natives happened to settle there. The two first mentioned are essentially railroad camps that have persisted since 1851, when they were successively the terminus of the road. Ahorca Lagarto, however, is on a bend in the river, and may well have been a resting place for

the cramped travelers in canoes. Of the origin of its name Otis says: "Ahorca Lagarto, 'to hang the lizard,' deriving its name from a landing place on the Chagres nearby; this again, named from having, years back, been pitched upon as an encampment by a body of government troops, who suspended from a tree their banner, on which was a lizard, the insignia of the Order of Santiago." In 1908 it had sixty-two inhabitants.

Bohio appears to have been another bush hamlet in 1862 when Otis wrote. Until recently it has been called Bohio Soldado (Soldier's Home). The French made it the site of one of their district headquarters in 1862, erected a machine stop on the west bank of the river, and did considerable work there under the old sea level plan for a canal,

which was excavated to this place to a sufficient depth for light draft boats. Here, as well as at any place can be seen today the plan of the sea level canal, which included the main channel and two large diversions or drainage ditches one on each side of the canal proper.

Under the French plan for a lock canal, Bohio was the site of the first dam, and the excavation for the locks at this point can be seen in one of the hills on the opposite side of the river from the railroad. As it has existed during the American regime the village has been a relic from the French period. Such surveys, investigations, and excavation as were necessary here were done by men occupying the French houses. In recent years Bohio has been the center of a small local trade in vegetables, brought in from the jungle by canoe and pack animals, in exchange for groceries and liquors sold in the Chinese and native shops. At the time of the official census in 1908, it had 526 inhabitants.

At Bohio the Americans carried on investigations in 1904 and 1905 to determine whether that location could be used for locks and a dam, and in 1909 excavation by hand and with steamshovel was carried on to remove a small hill and part of a dump made by the French, which stood in the canal prism. Across the river, where the machine shops were situated in the French days, and where they carried on work for the lock emplacement, the edge of a hill is now being removed by a contractor. The work at this point is typical of all that between Gatun and Culebra Cut, consisting as it does of the excavation of small elevations in the Canal channel and the toes of hills that project into the prism.

Post Office at Tabernilla, CZ.
Postcard courtesy of Robert Karrer.

As the Chagres River rose, the town of Tabernilla disappeared under Gatun Lake.
Postcard courtesy of Robert Karrer.

Canal de Panamá: Una vista de Tabernilla durante la inundación. A View of Tabernilla during flood.

Canal de Panamá: El Campamento de los trabajadores inundado. Laborer's barracks flooded at Tabernilla.

Postcard courtesy of Robert Karrer.

Historical account paints picturesque scenes of life in construction towns

The Panama Canal Spillway, August 9, 1996, vol. XXXIV, no. 16, p. 5.

Part 5

Frijoles

Known for such things as rum and bananas, the colorful village of Frijoles had also been a center for relocation work on the Panama Railroad.

Near Bohio are the hamlets of Penas Blancas and Buena Vista, both on the river and each merely a collection of huts of various descriptions.

Frijoles is the next railway station, a village of 784 inhabitants in 1908, of about a thousand when it became a center for relocation work on the Panama railroad, now being rapidly deserted. Here, for many years, an old Frenchman ran a distillery in which he made rum of such good quality that he boasted that it was sold in Colon to rectifiers who made it into "genuine French cognac."

One of the familiar sights of this hamlet is the village washing place—a pool near the railway tracks, formed by the swirling of the water in the Frijolita River at a point where it is turned at right angles to its previous course by the interposition of a bank of clay and rock. The method of washing clothes among the natives and West Indians can be observed here. This also is locally known as the place where one may buy bananas of peculiarly delicious flavor.

Frijoles is mentioned in F.N. Otis' guide book published in 1862, but the next village, Tabernilla, is not. It was one of the centers of the French work, and there was a small field repair shop at this point, with a few buildings that served as quarters for the working force.

During the American occupation it became a village of over two thousand inhabitants (2,079 in 1908), because here is situated the largest dumping ground on the canal work. The location was chosen in 1906

Gorgona

The lake village of Gorgona, either by design or coincidence, bore the same name that Francisco Pizarro gave to an island off the coast of Colombia after experiencing its treacherous currents.

because it is on the main line of the railroad and outside the canal prism and it afforded a plot of ground two miles long and almost as wide for wasting of spoil. In all, about sixteen million cubic yards of material were wasted here, all of which will be below the level of the lake.

The dump was abandoned at the close of 1910, and immediately the village population decreased, the people remaining there being largely employees with families who could not procure quarters elsewhere. These are now being moved because the demolition of the place is under way.

Between Tabernilla and San Pablo, the railroad crosses the Chagres River at Barbacoas. The original bridge was built of wood, but, early in the history of the railroad, it was replaced by a bridge of six wrought iron through plate girder spans, ranging from 101 to 109 feet in length, supported upon seven masonry piers.

This bridge is mentioned by Otis in 1862 and is said to have been one of the first of its type ever constructed. It was not built, however, to carry such heavy rolling stock as that placed on the road by the Americans, and so the three channel spans were replaced in

1908 by heavier girders, while the floor system of the three remaining spans of the old bridge was reinforced.

The Chagres River bridge at Barbacoas was said to be one of the first of its type.

San Pablo was originally a plantation worked by Catholic priests. It was a railroad station in 1862, was a laborer's camp in the French days, and, during the American occupation, has been a small canal village. It also is being demolished, and the last excavation in the lake region is now in progress there.

Across the Chagres River from San Pablo is Caimito. It was a canal labor camp in the French time and also under the Americans, until the work at that point was finished. Of this class, also, is Mamei, likewise a railroad station in 1862 and little more than that today, although it was the location of several quarters for canal workers a few years ago.

Gorgona bears the name given by Spanish conquistador Francisco Pizarro to an island off the coast of Colombia near Buenaventura, because he found around it such treacherous currents. It may be that this name was adopted arbitrarily, or that the Chagres River travelers found in the river at this place some eddies that reminded them of the currents at Gorgona Island.

Of this place Otis says, "The native town of Gorgona was noted in the earlier days of the river travel as the place where the wet and jaded traveler was accustomed to worry out the night on a rawhide, exposed to the insects and the rain, and in the morning if he was fortunate regale himself on jerked beef and plantains." In the French time, large shops were situated here at the point where the American shops now are, known as Bas Matachín.

Gorgona should not be classed with Gatun and Bohio as a purely jungle hamlet because it appears to have been a settlement of some size long before the railroad was built. It was one of the places at which river travelers stopped for the night, and all about it were cultivated farms. At the time of the first Canal Zone census in 1908, its inhabitants numbered 2,750. The population has increased owing to the expansion of work in the shops.

The site of the shops and the lower parts of the village will be covered by the water of Gatun Lake, and, therefore, the shops will be moved in about a year to the site reserved for the permanent marine shops at Balboa.

Canal Record announces fate of lake trees

The historical series about the villages that were covered when Gatun Lake was flooded may have caused some readers to wonder what happened to the trees in the area. This article, reprinted from the April 24, 1912, Canal Record provides some enlightenment.

Elsewhere in this issue of The Canal Record will be found an advertisement asking for bids for the privilege of cutting timber in the area within the Canal Zone that will be covered by Gatun Lake. The conditions governing the proposed contract are as follows:

(1) The privilege consists of a right to cut timber within the defined area until such time as the timber may be required by the Government, in which event the privilege would cease as of date of due notification to that effect to the contractor.

(2) The privilege would under no circumstance extend beyond January 1, 1915.

(3) Bidders will submit bid in one lump sum for the privilege.

(4) Payment of the full amount of bid will be due and payable within thirty days from notification by the Chief Quartermaster that bid has been accepted.

(5) Upon payment of full amount of bid the successful bidder may proceed at once to cut and remove timber within the area designated and may continue to cut and remove same until notified by the Government that

Tabernilla
Before it was covered by the waters of Gatun Lake, the village of Tabernilla contained the largest dumping ground for spoil material along the Panama Canal construction route.

the privilege is recalled, or until the first day of January, 1915.

(6) The successful bidder will be required to remove all timber cut by him and to make disposition by burning or otherwise of all brush and limbs resulting from such cutting.

It is not known how much desirable timber there may be in the area to be flooded within the Canal Zone. It is probable there is only a small amount of hard woods, because the land along the Chagres River has been occupied to some extent for three hundred and fifty years, and most of the original wood has been cut, while much of the area has, been burned over many times. So little desirable timber was found along the route of the Panama railroad at the time of its construction, 1850-55, that most of the ties were brought from Colombia. During the dry season of 1910 about 842 acres of land in the ship channel through the lake were cleared by contract, and in this area there were found only a few hardwood sticks and a small number of espeve trees, a wood with little grain, from which 17 native canoes were made. The contractor obtained about 4,000 bags of charcoal at 40 pounds to the bag, and this was the most valuable by-product of his work. Among the cabinet woods found in the lake region, a few sticks here and there, are cocobolo, guayacan, Panama mahogany, Spanish cedar, and several light colored woods that are very hard and take a high polish. Many of these woods are too heavy to float.

Typical American Family Quarters in the Canal Zone

NEW CONDITIONS OF EMPLOYMENT.

Canal Record, vol. I, no. 6, October 9, 1907.

OLD REGULATIONS
Effective April 1, 1907.

No application for married quarters will be received until an employee has been in the service on the Isthmus for six months.

NEW REGULATIONS
Effective July 1, 1907.

Family quarters will be assigned when available, assignments to be made in accordance with date of application. Experience shows that about ten months elapse between application and assignment.

Top two photos: A family enjoys their screened-in porch in San Pablo, CZ.

Left: Mother and son stand outside their home in San Pablo, CZ. Notice the shotgun leaning against the bag, probably a necessity for living in jungle. *Postcards courtesy of Robert Karrer.*

Right, Once the Panama Canal Zone was freed of mosquito carrying diseases, American workers were encouraged to bring their families to live with them in the airy screened-in houses built by the Panama Canal Company.

High ceilings, large windows, and wicker furnishings gave the homes an airy feeling, despite the fact that there were no air-conditioners in the tropics.

Photo from *Panama and the Canal*, Alfred B. Hall, 1910.

Additional Family Quarters

Canal Record, vol. I, no. 6, October 9, 1907.

On October 1, the number of applications for family quarters in excess of present accommodations, completed, under construction or authorized, was 309, as follows:

East La Boca 26	Las Cascadas 21
Ancon Hospital 5	Gorgona 53
Ancon 15	Tabernilla 5
Corozal 2	Bas Obispo 3
Pedro Miguel 16	San Pablo 1
Paraiso 13	Gatun 30
Culebra 33	Colon 39
Empire 34	Cristobal 50

All these applications have been filed since the beginning of the present fiscal year and the construction of the desired quarters, including furniture, light, water, and sewer system extensions, sidewalks, etc., would entail an expenditure of fully $900,000.

In view of the additional applications which are constantly being received, it is not believed advisable to incur this large expense without ascertaining fully the attitude of Congress, as practically all this expense would have to be covered by a deficiency - appropriation - this year's appropriations being insufficient for any of the work.

No action, therefore, will be taken toward approving or disapproving the construction of these additional quarters until after the visit to the Isthmus in November of the Congressional Committee on Appropriations.

A young couple relaxing on the screened-in porch of a Canal Zone home. *Courtesy of Malcolm Stone.*

Interior view of a married quarters.

A Steam Shovel Man's View

From a letter in the *Steam Shovel and Dredge*, August 1907.
Republished in *The Canal Record*, September 18, 1907, vol. 1, no. 3, p. 7.

We live in nicely furnished rooms with baths, electric light, and toilet rooms, and the board is exceptionally good. In fact, everything is done to make it as pleasant as possible for the men, and I have not seen a man that was not satisfied; as for myself, I like it very much. This is a pretty town. I am stationed at Culebra, which is the largest and deepest cut on the Canal. It is forty miles south of Colon and is only eight miles from Panama City, which is the Pacific entrance to the Canal. There are twenty-eight shovels in this one cut. It is a great sight to see so many working in one place at the same time. Some of the shovels are 200 feet above the others The first week I was here I was on a shovel that was on the top lift, and you could look down the smokestacks of a dozen other shovels below you.

Culebra has a population of about 2,000 people—all government employees. The town is laid off in terraces and graded with streets and sidewalks. The buildings are large and new, with mosquito netting from top to bottom. However, they do not need the netting, as the mosquitoes have been killed a year ago. I have not seen one since I came. The nights are cool—in fact, you have to sleep under a blanket. The days, however, are very warm.

A great many of the steam shovel men have their families with them, and if I stay on this job I intend to have my family come here just as soon as I can get a house. There is a great demand for houses here, and it is hardly probable that I can get one for several months. At the present time they are building houses on two new streets. These houses would rent for $50 per month in most of the cities in the States. These are the houses they give the married men, fitted up nicely with baths and electric lights, free of charge.

Things in the commissary are about as cheap as they are in the States. Everything in the commissary has to pass an inspection, so as a general thing, you get good stuff out of it. We have a Local Lodge, No. 19, of the International Brotherhood of Steam Shovel and Dredge Men, which has elected Mr. Bates as president.

Fraternally yours,

- A BROTHERHOOD MEMBER

Accidental Explosion

The Canal Record, November 20, 1907, p. 90.

An accidental explosion in the Cut on Friday last resulted in the death of two colored laborers and came near to causing loss of life in the town of Culebra. A small charge of powder had been placed unknowingly near an unexploded charge of dynamite of about 250 pounds. When the former was fired the latter exploded, and as it was only 12 feet below the surface in a stratum of soft rock, a large quantity of the overlaying material was shattered and thrown to great distances in the air, principally in the direction of the town of Culebra. Fragments of rock fell upon a number of buildings including the Administration Building, the Commission club house, the Culebra hotel, the Commissary building and various family and bachelor quarters. Some of the heavier pieces passed entirely through the roofs, and as a result there were several narrow escapes on the part of the occupants.

American Bachelor Quarters and Recreational Facilities in Canal Zone

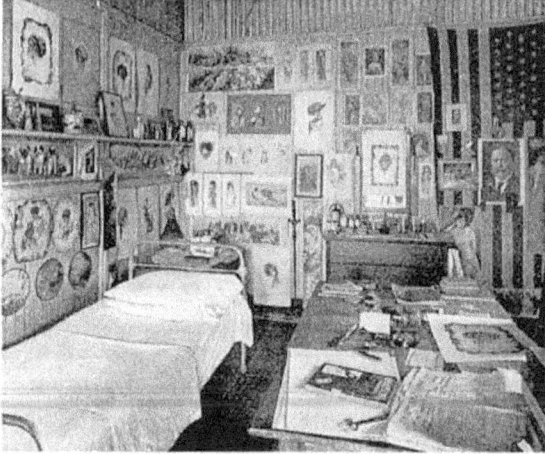

This was described as a typical room in the Bachelor's Quarters from *Panama and the Canal*, p. 334.

Bowling Alley from *Makers of the Panama Canal*, p. 168.

The Isthmian Canal Commission had clubhouses built in larger Canal Zone towns operated by the Y.M.C.A. for the benefit of employees.

Billiard Hall from *Makers of the Panama Canal*, p. 161.

Reading Room from *Makers of the Panama Canal*, p. 156.

Bachelors enjoying themselves in a group photo.
Courtesy of Ted and Patsy Norris.

Pay and Privileges

The Canal Record, December 4, 1907, p. 106.

The matter of quarters is not deserving of serious argument. The Isthmian Canal Commission is compelled to provide quarters for its employees because there is no alternative. Apart from buildings bought or constructed by the Commission there is scarcely a habitable house between Panama and Colon. This applies to married quarters as well as bachelors quarters. The Commission should scarcely recruit its employees exclusively among single men, or make celibacy a condition to continued employment. So long as married men are employed—and their employment is necessary—quarters must be provided for them.

Letters From the Line

The Canal Record, March 11, 1908, p. 219.

I read with amusement the suggestions offered in your valuable paper to remedy the so-called existing evils in the bachelor quarters. It appears to me that these evils are greatly exaggerated by these very good men, as there are certainly times when a little merriment is indulged in, but this God-given right to amuse ourselves tends in my estimation to help relieve the monotony of life on the Zone, and who will begrudge these little outbursts of gaiety; but a crank who in all probability has never left home until he arrived on the Isthmus? I find the occupants of the bachelor quarters the finest body of well conducted men to be found anywhere. The only suggestion I can make is, that a house be built away in the jungle, where pious people may sit out on the veranda and listen to the musical croak of the frogs.
- *Vincent Leat.Gorgona, March 6, 1908*

Ancon

Sponsors

Lester, "Chefa," Gladys & Lena Barrows
Harry, Mary, Katherine & Billy Egolf
Charles W. "Chuck" Hummer
Ernest "Buck" Krueger
Nathaniel, Marie, Naomi & Ruth Litvin
Michael & Elaine Stephenson Family
Family of James & Stacia Walsh
Joe Wood

Your Town - Ancon

The Panama Canal Review, October 1, 1954, pp. 8-9, 11.

THE OLD TIVOLI makes a background for some of Ancon's newest houses. President Theodore Roosevelt and his official party were the Tivoli's first guests.

If the powers that were in 1904 and 1905 had been able to figure out an inexpensive and practical method of reaching the top of Ancon Hill, Ancon would not look anything like it does today. Glowing from the summit of the 654-foot hill would be the lights of homes and hospitals instead of today's constellation of aircraft warning lights.

For months after the United States took over the rights and properties of the French Canal Company in 1904, American officials discussed some feasible way of reaching the hilltop. They agreed that it was a perfect location for anywhere from 14 to 25 houses and two sanitariums.

They considered and discarded the idea of a regular railway, a cog railway, a cable railway, and a macadam road "with a stage running at frequent intervals." The plans which were suitable to the steep ascent were unsuitable because of cost. Finally, in April 1905, they abandoned all idea of the summit site and settled on the lower slopes of the hill for their town.

Ancon: Anchorage

The name Ancon, which means roadstead or anchorage, goes back hundreds of years in Isthmian history. In 1545 Gonzalo Pizarro, seeking to control the Isthmus of Panama and its rich ports, sent two expeditions from Peru. The first pillaged the old city of Panama before it was recalled. The second was divided into two forces, one of which, under Rodrigo de Carbajal, landed at Ancon, a small cove two leagues from Panama.

In 1674 the new city of Panama was laid out beside this cove; 200 years later the French Company selected a hillside overlooking the roadstead as the site of its hospital. When the Americans came they used the name "Port of Ancon" for what was later to be the Pacific terminus of the Canal.

The first town of Ancon was one of five Canal Zone municipalities, each administered by a mayor and council. According to the Isthmian Canal Commission, it was to be "the seat of the government of the Canal Zone and the place of residence of a large proportion of the Americans on the Isthmus." Although headquarters for the Canal construction force were moved to Culebra three years later, Ancon did remain the main governmental and medical center for the Canal Zone throughout the construction period. In 1912 a local writer commented that there were people in Ancon who had never seen the Canal construction except from the windows of railroad cars.

Hospital Grounds

The first American construction in Ancon was the repair and expansion of the rambling 500-bed French hospital. Some of its wooden wards dated back to 1883; they were reconditioned and pavilions and second floors added. By 1907 Ancon Hospital had 96 buildings, 47 of them wards.

All of these were in what was known as the Hospital Grounds. A gate across its palm-bordered entrance road, approximately opposite the present residence of the Episcopal bishop, separated the hospital grounds from the rest of Ancon. It was kept locked at night and late comers had to ring a bell for admission.

The first housing was crude. Married officers detailed to the Ancon police station lived in tents. One three-family house had only one bathroom. New quarters were built or old French buildings made fit for habitation as fast as possible and by 1908 Ancon was a village of 1,508. At that, it was less than a third the size of Culebra.

Outside of the hospital, the biggest buildings in town were the Hotel Tivoli, which had been opened officially in January 1907, and the Administration Building, now the District Court. After 1908, it housed the offices of the Civil Administration and of Sanitation, which until that time had been in Panama City.

Ancon In 1907

A supreme court and a circuit court were located near the present post office in an old French building, and a corral for the horses and mules which pulled official transportation was close by. Officials of the Isthmian Canal Commission were housed in large quarters on Fourth of July Avenue; the quarters once occupied by Joseph Bucklin Bishop, the Commission's Secretary, are now the residence of the District Judge. The quarters of the Chief Health Officer were on a knoll behind St. Luke's Cathedral, then a steepled wooden chapel; other doctors and officers of the civil administration lived in the large houses still standing on Columbia Road.

A frame building on Reservoir Hill, levelled two years ago when the present housing was laid out, was the elementary school. One of the teachers in that school was Mrs. Ora Ewing, now housemother for the junior college dormitory.

No Commissary

She lived in a four-family house on Fourth of July Avenue, looking out over a large field where the National Institute now stands. There was no commissary in Ancon, she recalls. Each week she made out food orders for six days and sent them and a $15 commissary book to Cristobal. The orders were delivered each morning after the train arrived from the Atlantic side.

ALL DRESSED UP, with someplace to go! The ladies donned their fanciest hats, with veils, and the gentlemen their most dazzling white flannels when the Ancon Amusement Association chartered the *SS Aysan* in February 1908 for a trip to the Pearl Islands.

HARRY CORN
In charge of Ancon's Post Office

S. D. CALLENDER
Pacific Service Center Manager

CHARLES L. LATHAM, Jr.
Ancon Commissary Manager

Unlike most other towns, there was no clubhouse in Ancon and therefore, no planned amusement. The people made their own fun. The men went to the various lodges, like the Masons, Kangaroos, or the Knights of Pythias, which had space over a quartermaster storehouse; the women belonged to the very active Woman's Club, and later, to the Ancon Morning Musicale Society, disbanded only a few years ago.

The Tivoli was the scene of regular cotillions; the Ancon Amusement Association, of which Mr. Ewing was a charter member, arranged dances, picnics, sports, and even, once, chartered a large ocean-going steamer for a trip to the Pearl Islands. Round trip was $2 a person; children under six went free.

Society was fairly formal. Calling cards were part of every lady's equipment, and oldtimers remember one woman who always wore a hat to her own tea parties.

Permanent Town

By 1907 Ancon was beginning to spread beyond its original confines of Hospital Grounds and Tivoli. Fourteen buildings were constructed in what is now the San Juan Place area to house the insane patients who had been cared for originally at the hospital proper and later in a building near Miraflores. About 1915 the patients were transferred to Corozal and the buildings converted to quarters. For years they enjoyed the dubious distinction of being haunted by crazy ghosts which, anyone knows, are worse than regular ghosts.

In 1910 the corral was moved to a location not far from the Insane Asylum where it remained until 1938 when it was moved to Gaillard Highway.

As the Canal neared completion, Ancon's future was uncertain. Finally, in 1914 THE CANAL RECORD announced the official decision that the settlement at Ancon be continued indefinitely. The permanent force to be housed there was to be about 161 families and 130 bachelors. A number of quarters were brought in from Culebra, Empire, and Bas Obispo and re-erected in Ancon and some new houses were built.

Construction Began

The present commissary was built in 1914 on the site of an old French building; for years tradition had it linked by a tunnel to the hospital buildings. Plans were drawn up for a new school, still in use, and a clubhouse, the town's first, was built near the commissary. This clubhouse, a two-story frame structure, was burned to the ground in 1924 in Ancon's most spectacular fire.

Capt. R. E. Wood, Chief Quartermaster,

Dr. A. B. Herrick, then Acting Superintendent of Ancon Hospital, and Samuel Hitt, Canal architect, were appointed to a committee to submit recommendations for reconstructing the hospital. The first of the hospital's present buildings was authorized in 1915. The Ancon restaurant, later known as the Clubhouse, was built in 1917.

Up to that time the streets had no official names. Houses were numbered, and not very logically at that, and locating a particular residence was difficult. C. A. McIlvaine, then Executive Secretary, suggested a number of somewhat poetic names--Lovers Lane, High and Low Roads, Sleepy Hollow, and Palm Court--but his suggestions were not adopted. It was not until 1920 that Ancon Boulevard, Gorgas Road, Columbia and Culebra Roads, for instance, came into official being.

For a decade the town went on its quiet way. Its only excitement in the early 1920's was a rash of burglaries committed by a daring character named Peter Williams. Old timers declare that he used to notify the police in advance of the location of his next burglary. Whether or not that is true, he eventually fell afoul of the law and was shot and killed by the police while fleeing along an Ancon drainage ditch.

Fishbowl, Doctors' Knob

By the early 1930's the Ancon housing situation was acute; newcomers had no place to live except "vacation quarters" for sometimes as long as two years. Twelve two-family houses and two cottages were built near the hospital for doctors' quarters. The section was immediately and obviously christened the "Fishbowl."

The early 40's brought the completion of the houses in the Old Corral area;

emergency 12-family housing on Frangipani Street and new quarters for the doctors on Herrick Heights. Unnamed for some time, this section near the Ancon Courthouse was irreverently referred to in official files as "Doctor's Knob."

Then came the war. Like other towns, Ancon lived through blackouts, air-raid alerts, civil defense drills. Makeshift sandbag shelters appeared under houses and in side yards; starred service flags swung in windows. Planes from North and South America landed and took off from the Air Terminal on Gaillard Highway, now the Civil Affairs Building. Almost overnight a huge building appeared over the edge of the hill below the Tivoli; built by the USO for use of the thousands of local and visiting servicemen, it is now the Pacific Clubhouse.

After the war, building was again resumed in Ancon. About 1950 the new "Gyn-Ob" building was added to the Gorgas hospital group, and a site cleared for a clinic building, still to be built.

Flame Throwers

The haunted old quarters and the beautiful trees on San Juan Place fell before the bulldozers. Reservoir Hill, for many years a residence for women bachelors was so levelled that it can no longer be called a hill. Ancon Boulevard was relocated and the dreary old four-family houses which had stood flush with its sidewalks became debris. Army flamethrowers burned some of the wreckage where it lay.

Today some parts of Ancon look like the gap-toothed mouths of little boys. Old houses are being torn down as they are vacated, and it will not be many months before they are all gone.

Boundaries

Where Ancon begins and ends is

MISS DOVA ANTILL
School Principal

DR. ROBERT BERGER
Ancon's Doctor

a matter of question. On one side it is separated from Panama by Fourth of July Avenue, whose name dates back at least to 1909 and probably earlier. The border, generally, is the Panama-side curb of the avenue. On the other end of town the dividing line between Ancon and Balboa, for taxi-fare purposes, is the Administration Building. For school purposes, however, the Fishbowl, San Juan Place, and the Old Corral area, and everything west of these sections, are considered in the Balboa elementary school area.

It has three churches, all well patronized. Sacred Heart Chapel and the Cathedral of St. Luke began their lives as chapels inside the hospital grounds. The First Church of Christ, Scientist, on Ancon Boulevard was once the Ancon courthouse. It was moved to its present location in 1916 and has since been enlarged.

Voracious Deer

Ancon has a greenhouse, the only one in the Canal Zone. It has the only hotel in the Canal Zone, though it is now known as the Tivoli Guest House. Its hospital is the only one on the Pacific side. The marriages,

General View of Ancon, showing Administration Building, Panama

adoptions, divorces, and estates of hundreds of its citizens are recorded in the files of its District Court.

It also has some of the most voracious deer outside a jungle. One resident has an electrified fence around his yard to keep them out of his garden.

Many of the people who live in Ancon have never lived anywhere else. Nor do they want to. Once an Anconite, always an Anconite, they'll tell you.

Old Houses in Ancon Scheduled For Razing Early in Coming Year

The Panama Canal Review, March 5, 1954, p. 1.

One of the most popular of the new on-the-ground houses, is Type 337, popularly known as the "mother-in-law house."

The opening of bids this week for the construction of seventeen Type 337 houses in the Ridge Road area of Balboa Heights was one of the final steps to be taken in the Canal's long-range housing replacement program. The new houses to be built in this area are the type which has proved to be one of the most popular of any erected in the long-range replacement program. This type has been popularly designated as the "mother-in-law" house because one bedroom and bath are located across the patio from the other two bedrooms.

Next year's program calls for the replacement of the older quarters in the Ancon-Balboa area with 40 family apartments. All of the 1907-1914 houses on seven streets in Ancon are scheduled for demolition.

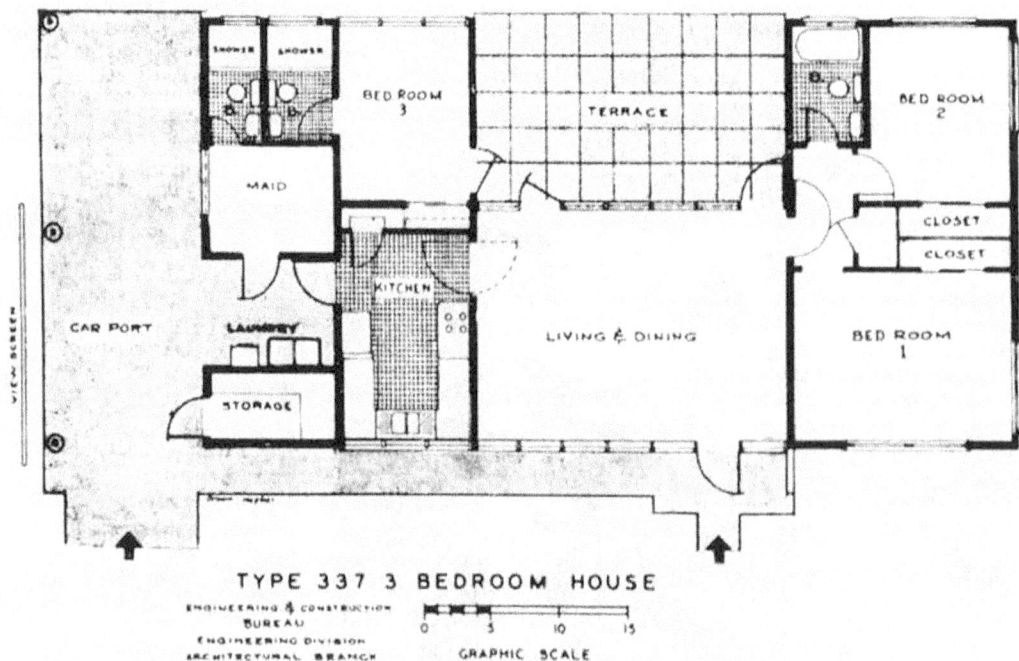

TYPE 337 3 BEDROOM HOUSE

ENGINEERING & CONSTRUCTION BUREAU
ENGINEERING DIVISION
ARCHITECTURAL BRANCH

GRAPHIC SCALE

After Nearly Half a Century

The Panama Canal Review, March 5, 1954, p. 11.

Another Canal Zone Landmark is nearing the end of its life. The "Admiral's House" on Fourth of July Avenue near the District Court is being vacated. March 31,1954, is the date slated for the last occupants to move to other quarters. After that date the old building, which was first occupied in 1907,

will probably be demolished.

It is probable, although old records fail to show it, that the first occupant of the big building was J. C. S. Blackburn, who headed the Civil Administration Department and was a member of the Isthmian Canal Commission until 1908. At the end of the construction

period the house was used for several months by the Joint Land Commission. After 1933, the building, became a women's bachelors quarters and was assigned to a group of teachers who ran the house as a bachelor "mess" for several years. In 1942 it was converted into two apartments.

Just Ask The Staff Of The C.Z. Library

The Panama Canal Review, September 3, 1954, p. 5.

A not so large room in the Administration Building at Balboa Heights was the library's original location 40 years ago. Today there are more than 103,000 volumes on the shelves of the main library on Gaillard Highway or distributed among the library's two branches at the Cristobal and La Boca and five deposit libraries at Gamboa,

Gatun, Paraiso, Santa Cruz, and Margarita. The original library's collection was made up mostly of general engineering books and pamphlets. There were no books of fiction, travel, or biography for readers except what was available at the YMCA clubhouse. A committee was formed in 1916 to see what could be done to improve the situ-

ation. By 1924, the library had almost 6,000 registered borrowers, and its collection increased to over 20,000 volumes and almost 8,000 pamphlets. The library was part of the Panama Canal's Record Bureau until it became part of the Civil Affairs Bureau in 1950.

Original Post Office To Cease Operations

The Panama Canal Review, October 5, 1956, p. 13.

The closing of the Ancon Post Office on January 1, announced last month by Gov. W. E. Potter at the "town meeting" held at Diablo Heights, reduces to three the number of Canal Zone post offices which have maintained uninter-

rupted operations since June 24, 1904. The surviving oldtimers, after January 1, will be Cristobal, Gatun, and Balboa post offices. Cristobal and Gatun have operated since they were opened under those names; Balboa post office was

known as La Boca until May 5, 1909. No decision has yet been reached as to what use will be made of the Ancon post office building after postal operations are stopped there.

Seventy Five Years Of Medical Service

The Panama Canal Review, November 1, 1957—Diamond Jubilee Supplement.

"French Days - L'Hopital Central du Panama 1882-1904," p. 2.

On a September Sunday 75 yrs ago, Bishop Telesforo Paul of Panama celebrated a Pontifical Mass, giving thanks to God that the great new hos-

pital of the French Canal Company at Ancon had become a reality.

For those days 75 years ago, the hospital was an impressive one. The

best contemporary description of it appears in the Star & Herald of September 13, 1882.

"Construction Days - Ancon Hospital, 1904-1914," pp. 7, 10-11.

The first American doctors and nurses had their work cut out for them, and there was plenty of it to do. Three months earlier, when Colonel Gorgas came to Panama as a medical advisor with Isthmian Canal Commission, he had looked over the French hospital at Ancon and Colon. He had decided that both establishments could be reconditioned as a nucleus for an American hospital system, although he realized that the hospital buildings were in sad condition.

From the very beginning Ancon Hospital had the benefit of the best medical talent of those days. Dr. John Ross, a Navy doctor who became director of the hospital system under the Americans; Dr. Henry Rose Carter of the U.S. Public Health Service; and Major La Garde, Ancon Hospital's first superintendant, were all authorities on yellow fever.

Ancon Hospital, Section "C."

The Transformation - Ancon Into Gorgas, 1914-1941," pp. 12-13.

Even before the Panama Canal was completed and the construction period ended, the medical men in the Canal Zone were turning their thoughts toward a permanent hospital at Ancon. By April 1919 the entire hospital had been rebuilt, including a new home. The entire reconstruction cost $2,000,000.00.

Ancon Hospital was not only the largest medical institution in the western hemisphere south of the United States but it was also the only hospital in the area which could handle any type of medical or surgical cases.

As the fame of the big hospital grew and spread, friends who had long sought a way to honor the Canal Zone's first Chief Sanitary Officer saw it as a fitting way to preserve his memory.

Congress agreed and on March 24, 1928, passed a Joint Resolution which said: "In recognition of his distinguished service to humanity and as a fitting perpetuation of the name and memory of Major General William Crawford Gorgas, the Government Hospital heretofore known as the Ancon hospital shall hereafter be known and designated on the public records as the Gorgas Hospital."

"War and Postwar - Gorgas Hospital, 1941 and Since," p. 15.

Today, Gorgas Hospital is not only a teaching hospital—a status it has held for years—but it is a fully accredited medical institution. It is was so rated in 1954 by the Joint Commission on Accreditation of Hospitals, a body made up of the American College of Surgeons, the American Hospital Association, and the Canadian Medical Association.

Gorgas Hospital, in 1957, has 25 clinics and 37 administrative and special departments. Its staff of 716 includes 37 staff doctors, 16 residents, 15 interns, and 146 nurses.

It was a great hospital in 1882. It was a great hospital in 1904. It was a great hospital during two great wars.

It is still a great hospital today.

History Goes Up in Flames

The Panama Canal Review, March 4, 1960, p. 20.

Another bit of Canal Zone history came to an end last month. This time the demise was not due to the hammers and levers of wreckers, but to a fire deliberately started by the Canal Zone Fire Division.

This somewhat spectacular end was the fate of House 269 in Ancon, known to everyone who was once a child on the Pacific side--and to a good many adults as well--as the Birdwoman's House. Occupied for over 40 years by Mrs. Lucille Bryan and, before his death, A. H. Bryan, it had once housed one of the most extensive private collections of birds ever seen in these parts.

Page in Canal History Closed

The Panama Canal Review, May 5, 1961, p. 15.

Another page in Canal history was closed near the end of April as the last family to occupy house No. 364 in Ancon moved out, leaving the sprawling four-family building to the demolition crew which will tear it down to end a career started with its construction in 1907. The house is the last one in Ancon which was built prior to the opening of the waterway.

 Memories

Ancon, CZ

Ancon was unique among Canal Zone towns. Known as the gateway to Panama City, it was an exciting town, juxtapositioned on the CZ-Panama boundary near famous Panama landmarks such as Morrison's, the Kool Spot, Shaw's, the Linen House, Ricardo's Jewelers, the Ancon Inn, the Good Neighbor Bar, the Riviera Bar, Automobile Row and 4th of July Avenue, where picturesque lottery vendors lined the sidewalks "just across the street" in Panama to sell their tickets because their sale was illegal in the Canal Zone.

Zone residents would park their cars in Ancon and venture across the border, not only to buy lottery, but to go to Central Avenue via "J" Street where they frequented the Hindu stores, Nat Mendez' Jewelry store, Chinese restaurants, the Napoli, the Central, Cecilia or Tropical theaters and many of the bars that lined those streets, such as Hancock's, Kelly's Ritz and Happyland.

Ancon also had the magnificent Tivoli Hotel, which hosted many weddings, formal balls, high school proms, retirement parties and other important events. Who could forget the famous New Year's dances at the Tivoli, pulsating with the spirited music of the incomparable Lucho Azcarraga and his "conjunto," while the ballroom floor shook as people stomped to the beat of "Pescao!" and "God Bless America."

Ancon was also famous for Gorgas Hospital – the birthplace of many Zonians; the Ancon District Court; the Ancon Laundry; and, for many years, the Ancon Railroad Station, where the Panama Railroad began and ended its daily Coast-to-Coast journey.
- Joe Wood

Folklore Class Interview – Mrs. Olsen

When I first got down here in March, 1920, there were two towns – Ancon and Balboa. They were two different towns just like they are now except that there was more of a division then. Now they seem to blend together like one big city. For entertainment we could usually go to the movies. There were two movies a night – one in Ancon and one in Balboa. They timed it so you could go to one movie and then catch a bus and go over to the other one. We didn't have to use money to get into the movie if we didn't want to. We had books that we called Commy books which we used in place of money. There were Commy books for two dollars and fifty cents, five dollars and fifteen dollars. If you needed fifteen cents to get into the movie, then you would tear fifteen cents worth of tickets. In the commissary you couldn't use money; you had to use these books. We could even go to dinner and use our Commy book. The two dollar and fifty cent book was green; the five dollar book was white; and the fifteen dollar one was pink. The commissary was a wooden building just like many of the other buildings. Besides going to the theater we would spend a lot of time at the clubhouse. It was located in the same place that it is now except that it was an old wooden building then. The only brick buildings that they had when I first got here were the Masonic Temple, Administration Building, The Tivoli Hotel, the restaurant which was converted into the YMCA, the Catholic school (St. Mary's) and the Balboa Elementary School. All the rest of the buildings were the old wooden framed style. The oldest houses that I can think of that are still standing are a couple of gigantic houses on a hill in Diablo. They were built high enough so that they had a lovely view of the Canal, and to this day they are still being used and lived in.

Gorgas Hospital was running in full swing when I got here except it had the name of Ancon Hospital then. My first child's birth certificate says that he was born in Ancon Hospital and my second child's says that she was born in Gorgas Hospital – I had both of them in the same delivery room. They named the hospital Gorgas after the doctor who cleaned up the yellow fever in the Canal Zone.

In my earlier days here, every employee received a railroad book of tickets for the year. There were twelve to each book, which allowed us one free ride a month. Trains were at that time a very important means of transportation. The little Chiva buses were used more than the trains. These were little buses that had benches along the side, and you just crowded in to where you could find room. They would tie all their chickens and wicker baskets or whatever else they had on top or around the sides of those little buses.

Living in the Canal Zone gave me the opportunity to meet and see some famous people. I can remember when Lindbergh flew the first air mail flight down here in 1927. He landed in Panama somewhere and came to the Canal Zone. He went to the swimming pool, sat in the grandstands and watched the children perform. Mr. Bee was famous for his Red, White, and Blue troupe. They were a group of local kids that swam together. They were in the news and magazines and on television. Two of the young people also went as far as the Olympics. Admiral Byrd also stopped by on his way down to the South Pole. We were allowed to go aboard his vessels, and a friend of ours went with him on his travels. Ruth Elder was one of the first women that tried to go across the ocean in a plane. She didn't make it, but she wasn't killed either. Margot Fonteyn married an Arias, which was a very prominent Panamanian name. He brought her to Panama during the height of her popularity. Mrs. Gee, who was the dancing teacher in the Canal area at the time, was a very personal friend of mind and of Margot Fonteyn's.

My life in the Canal Zone has been wonderful. I made it my home and home for my family.
- *Sherry Elliott in 1978 for a folklore class taught by Mary Knapp at Balboa High School.*

Mango Tree Memories

As with most kids in the Canal Zone, we spent a lot of time outdoors. If we weren't exploring the jungle looking for insects and "Doodle Bugs," we were sliding in the mud when it rained or playing sports or games such as marbles, bottle-cap baseball, Ringo-levio, kick-the-can, hide-and-seek, etc.

We also played tree tag. One day, we were playing tag in a big mango tree near the Ancon Laundry. The object was to avoid being "tagged," or else you would be "it" and would remain "it" until you "tagged" someone else. While racing around the tree top, jumping from limb to limb to avoid getting "tagged," a branch broke and I fell about 20 feet to the ground. As they carried me off to the hospital, unconscious, my lifelong friend, Burt Mead, yelled to the other kids, "Hey, did anyone tag him on the way down?"

That tree was the source of our daily supply of green mangoes – which we ate with vinegar and salt – and also supplied the ammunition for our occasional mango fights.

Shade from that tree would cover the bleachers at the Pacific Softball League ballfield—called the "Beer League" for the daily keg of beer that was provided for players. Kids would hide behind the tree, sneak under the bleachers when the players took the field, grab some beer and run back to the tree. We never got busted, but the players knew what we were doing.
- *Joe Wood*

Ancon

Some of the games we'd play for fun were Ring-a-levio, Hide and Seek, and Kick the Can. We had boundaries for these games, but we always broke the rules. We made the red-neck kids look for us, and they could never find us because we always hid in people's houses. A couple of times I hid in the washing machine.

In Ancon there was this field where a lot of people would hang out. We played kickball, football, baseball, and battleball. We Girls were tomboys. We always played games with the boys. They had this big tree by the field, and some of the guys would tie a rope to the top of the tree; we would get our thrills swinging on it, which was pretty risky. We'd have war fights, dividing ourselves into groups by this round church in Ancon. One team would go on one side and the other team on the other side. We had ten minutes to pick up things to throw at each other. We used hot beans and also those hard round beans that fell from the trees.

The most popular thing we did was to slide down those steep hills in Ancon on cardboard. We'd have competitions to see who could slide the fastest down the hill.

One of our secret places was a hill up by Gorgas Hospital. It was like a jungle. There were a couple of little cement houses there which were empty. We'd have picnics up there. Another place was where the old Tivoli was. There was like a cellar, and at night we'd all get together and tell scary stories. We'd get really scared because the place was so freaky and scary. Another place we went to tell scary stories was in the jungle of the fishbowl.
- *Elizabeth Burgoon in 1978 for a folklore class taught by Mary Knapp at Balboa High School.*

Masonic Hall and Bachelor Quarters, Ancon

The Masonic Hall and Bachelor Quarters in the early part of the 20th century.

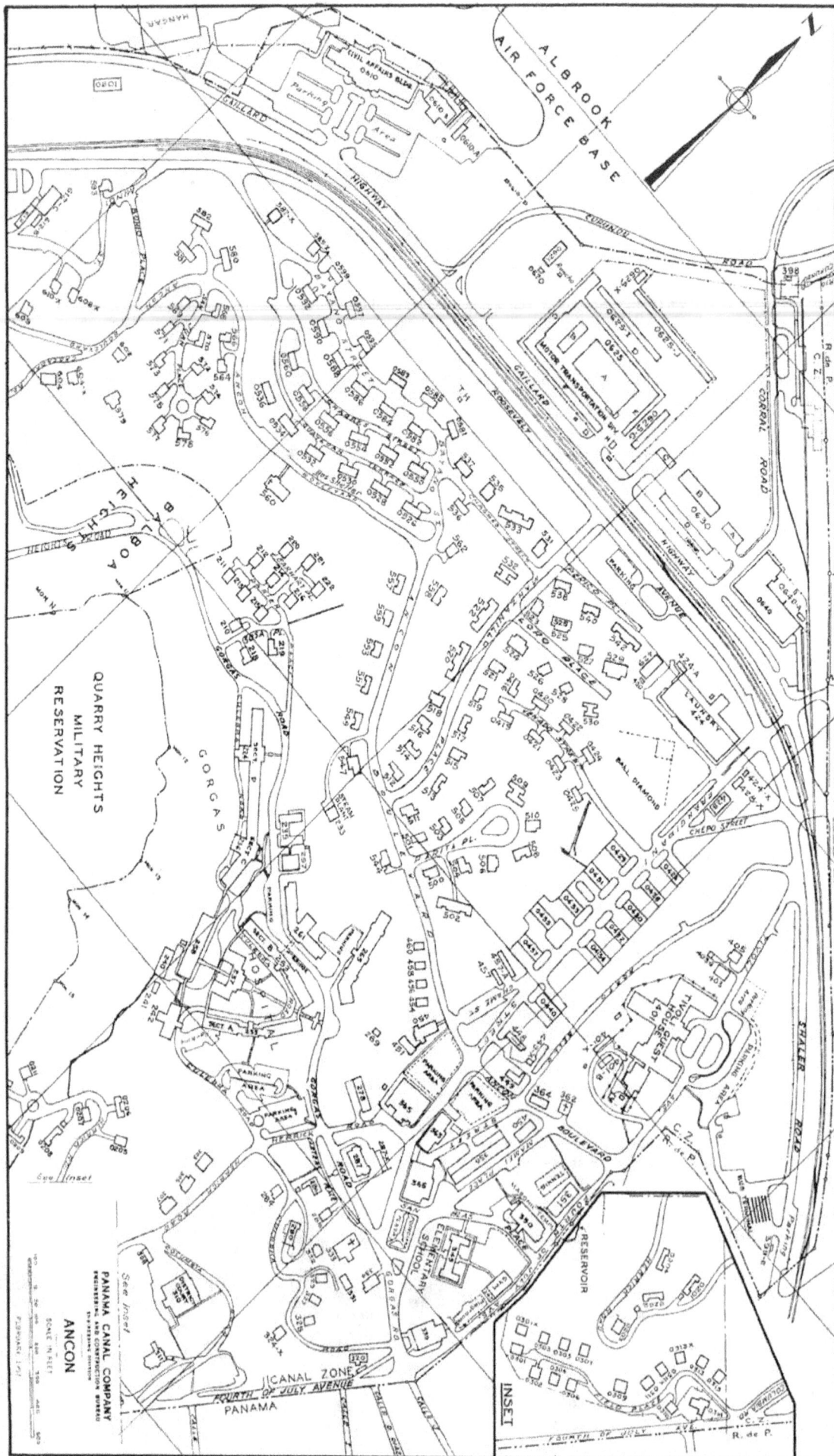

PANAMA CANAL COMPANY
ENGINEERING AND CONSTRUCTION BUREAU

ANCON

CANAL ZONE
PANAMA

QUARRY HEIGHTS
MILITARY
RESERVATION

ALBROOK
AIR FORCE BASE

INSET

26

Balboa
Balboa Heights
Sponsors

Baglien Kids: Lynn, Beth, Julie, & Joel
Lester, "Chefa," Gladys & Lena Barrows
Jack, Adelia, Sharon, & Jackie DeVore
W. R. Dunning Family, Bill, Pat, Danny, Sandi & Vicki
Harry, Mary, Katherine & Billy Egolf
Tamara Martinez Gramlich
The Hart Girls: Leona, Edna & Dorothy
Chuck Hummer
Nathaniel, Marie, Naomi & Ruth Litvin
Family of Paul & Margaret Morgan
Joan McCullough Ohman
Elma L., Ernestine & Fred Raines
Roy, Peggy, Royna, Jim & David Reece
Bea Monsanto Rhyne
Kaye & Leonor Richey Family
Michael & Elaine Stephenson Family
Roy, Geneva & Janet Stockham
The Lynn Stratford Family
Philip & Weulcia Wilkins Family
Joe, Bev, Craig, Brian & Scott Wood

Your Town - Balboa

The Panama Canal Review, July 1, 1955, pp. 8-9, 11

BALBOA, Pacific terminal town, was still in the early stages of construction when this photograph was taken December 23, 1914. The first offices in the Administration Building, on the Big Tree Hill, had been occupied a few months earlier.

Balboa, Pacific terminal town for the Panama Canal, could have been very different than it is today.

If the original Canal plans had been followed, there would be a lock about where the Balboa Terminal Building now stands and a great dam would link Sosa and Corozal hills; a man-made lake would cover what is now Balboa.

If town planners had had their way about 1913, the Balboa Commissary, Post Office, and Service Center would be connected by arcades which, in view of some November rainfalls, might not be a bad idea.

If studies made during the 1920's and 1930's had been carried out, the bulk of Balboa's housing would be located where La Boca today is fast disappearing.

And, if suggestions of a 1950 report had been adopted, Balboa today would be ringed by Morgan Avenue, which would start from the sea end of the Gavilan area, overpass Balboa Road, underpass the Administration Building's long front steps, Roosevelt Avenue, and the railroad tracks, and join eventually with Gaillard Highway.

But none of the "ifs" did happen. Today Balboa, second largest Canal town, is a sprawling assortment of houses, office buildings, schools, shops, churches, lodges, business establishments, and docks scattered over the flats, up and down hillsides and along the waterfront.

Named In 1909

The settlement on the Pacific end of the Canal was not called Balboa until 1909. The name was suggested by the Peruvian Minister to Panama who advanced the idea that the southern terminal should honor the discoverer of the Pacific, just as the northern terminal honored the discoverer of the new world.

Up to that time, the two Pacific side settlements in the Canal Zone—one extending from the old Panama Railroad Pier near what is now Pier 6 to the present Balboa docks area and the other a group of quarters and service buildings in the general area of today's Morgan Avenue-Pyle Street section—were known as Old La Boca and New, or East, La Boca respectively. After 1909 they became Old Balboa and East Balboa.

Old Balboa included a native town, which had once been known as Cartagenita, the marine shops and docks, a base for dredges working in the harbor, a few offices, and a restaurant with bachelor quarters on the upper floors. A few of the buildings dated back to the French days, as did the steel pier which was the Pacific terminus, a spur of the Panama Railroad.

Conveyor System

A sea of knee-deep, oozy mud frequently covered the area between the railroad tracks and the "hotel" which was on the side of a hill later removed to make room for the dry dock.

SUPERSALESMEN are these two Balboans: B.S. Chisholm, Manager of the Service Center, left, and J.F. Evans, Manager of the Commissary.

Morris Seeley, of Gamboa, remembers that this was the situation when he reported for work there in the fall of 1907. He was standing by the tracks, his small trunk-locker beside him wondering how he could reach the hotel, when he was approached by a huge Martiniquen laborer who offered in sign language to help him.

The Martiniquen put the trunk on his head and indicated to Mr. Seeley that he should climb on top of it. He could not be persuaded into making two trips, one with the trunk and the other with the man, so Mr. Seeley followed instructions, perched on top of the trunk and was conveyed safely to dry land.

East Balboa was on higher ground and was, primarily, a residential section for those working on the harbor development. La Folie Dingler—or Dingler's Folly—stood on the side of Ancon Hill above East Balboa, looking out over Panama Bay. It was a massive edifice which had been built in 1885 for but never occupied by, Jules Dingler, Director General of the French Canal Company. After the Canal Zone was established, it was used for several years as a quarantine detention station. It was razed in 1910.

Between the two Balboas were several community buildings. A "hotel," construction day term for a restaurant, was approximately opposite the present day St. Mary's School. The building was later used by the YMCA. The commissary, opened in 1909, the post office, which sold its first stamp on May 5, 1909, a lodge hall, still standing but soon to be demolished, and a one-story office building, housing the Assistant District Quartermaster, were clustered around what is now the Barneby Street-Balboa Road intersection.

Tide Waters

This "civic center" was separated from East Balboa by a wide ditch through which tide waters flowed twice a day as far as the base of the hill on which the Administration Building was later built. The ditch was crossed by an iron bridge. When the flats were filled with spoil from Balboa harbor, the bridge was buried; it was uncovered some years ago when La Boca Road was built.

Railroad tracks skirted the foot of Administration Building hill. One spur ran from Ancon to the old PRR pier, around the north side of Sosa Hill. Another set of tracks ran to a dump area—now Fort Amador—along present-day Amador Road and Empire Street.

Fred deV. Sill, a gay blade in those days, recalls that after a dance at the Tivoli, Isthmian beaux would "borrow" a railroad handcar and ride their ladies to the end of the line and back, pumping manfully. Even on a dry-season night, such exertion did a stiff collar no good.

By 1910 it was definite that there would be a large harbor at Balboa, and that a terminal town would be built near the Pacific entrance to the Canal. A huge harbor—it was first to cover 176 acres and was later enlarged to 246 acres—would provide a haven for ships. Typical of those happy-prenuclear days was this comment from THE CANAL RECORD:

"The harbor as planned will afford an absolutely safe anchorage. It will be practically landlocked and merchant

28

CAPT. B.A. DARDEN
Police District Commander

shipping will be secure from all attack as the harbor will be directly under the lee of the proposed fortifications on Sosa Hill."

Several groups planned layouts for the terminal town, but it was not until 1912 that a final decision was reached. Austin Lord, the architect who designed the Administration Building, laid out the town around a central avenue, now the Prado, with community buildings grouped at the Sosa Hill end connected by a continuous arcade. Quarters for the office force of the Administration Building were to be adjacent to and northeast of the Building. Employees of the shops, docks, and terminal facilities would live in quarters to be built on the slope of Sosa Hill and on the new fill, later to be known as The Flats.

The architect's plan did not meet universal approval. One dissenter objected on the grounds that there might be goings-on around a police station not fit for the sight of women and children and that, furthermore, "a space 74 by 22 feet for the sleeping of 46 prisoners and containing 26 plumbing fixtures is surely luxurious confinement." In addition he thought that a dormitory, instead of rooms in the individual quarters, should house the servants.

By late 1913, Balboa townsite was fairly well laid out. When the House of Representatives Appropriations Committee arrived in November they found white-painted stakes marking the street pattern.

Town Fills Up

Load after load of dirt, from Balboa harbor and Diablo Hill, were hauled to fill the flats and hydraulic dredges added their considerable bit. Meantime, work began on 10 four-family concrete quarters on the slope of Sosa Hill, the first to be built in Balboa. It was not until 1917 that the frame four-family houses in The Flats could be constructed.

By the latter part of 1914, Balboa had begun to fill up. Over 2,000 Pacific siders gathered on the piers to watch the SS Ancon complete the first formal Canal transit. By the following July, Balboa's residents numbered 893 men, 230 women, and 250 children. The children attended school in a building which had been put together from four frame quarters which had once housed bachelors. The high school, on the top floor, had a faculty of five, including its principal, Jessie H. Daniels. The eight elementary grades were on the first floor where Elise Cage, as principal, supervised her five teachers. Neither teachers nor pupils suffered any adverse effects from the police station-jail, directly across the street.

Clubhouse Opens

The clubhouse, once the pride of Empire, was moved to Balboa and officially opened on Christmas Eve 1914. Five months later Balboa commissary admitted its first customers who were duly impressed by the baskets of flowers which were the opening day gift of the landscape architect's gang.

At first, Balboa had nowhere nearly enough quarters to accommodate the families to be assigned there. For a while some Balboa families lived in La Boca. Closing of Corozal as a Canal town in 1915 threw an additional load on the already overstrained housing and it was several years before the situation eased.

Balboans, like other Zonians, made their own fun. One early event was a Christmas dance in the Old Balboa hotel given in 1911 by the Balboa volunteer fire department. In 1915, the town celebrated its first July Fourth. Fireworks went off as scheduled from Ancon Hill but a heavy storm rained out a street dance. The merry-makers took it in their stride, adjourning to the Balboa Clubhouse and the Tivoli.

Trams And Movies

There were tramcars to ride to Panama, for women market-bound, and far out on the Sabanas, for families on a Sunday outing. After automobiles became more numerous one could drive around "The Loop," through what is now Albrook Field, or along Amador Road to inspect the fast-growing new military post, or onto the docks.

When the Balboa restaurant, now the police station, was built about 1917, it became one of Balboa's social centers. Some people even dressed—in the society meaning of the word—to dine there and it was the fashionable place to drop in for coffee or hot milk after a dance.

Other recreation centers were the Balboa Community House, now St. Mary's School, the Balboa Yacht Club which today, remodeled, houses the American Legion, the YMCA, and a YWCA which was located on Carr Street and later became the dormitory for many students at the Junior College.

By 1920 the townspeople were

CAPT. WILLIAM E. JONES
Fire District Commander

practically busting out of the small movie hall on the second floor of the clubhouse and a special "moving picture room was added to take care of this increased patronage" according to the Governor's annual report. 1950 the new "Gyn-Ob" building was added to the Gorgas hospital group, and a site cleared for a clinic building, still to be built.

Pranks And Pranksters

One grownup Zonian still remembers with some delight, how he and his friend used to break up movie performances regularly by rolling stones down a drain which ran under the building. Which is probably as good a place as any to talk about some of the minor outrages perpetrated by Balboa's younger set.

The ringing of the school bell on Halloween was traditional for years; the students would go to any lengths to rig up the necessary Rube Goldberg apparatus. Another, more costly, prank involved throwing dye into the swimming pool. For years, Balboa small fry had a running feud with the firemen to see who could be first to burn the dried grass on Sosa Hill.

For several weeks one year, homeward-bound servants were terrified almost out of their wits by white-sheeted young Balboans who jumped from behind

THIS TRIO rules school-going Balboa. T.F. Holz, left, is principal of Balboa High School. Mrs. Marie Neal is principal of Balboa Elementary School. And Roger D. Michel keeps a firm hand on the subteens in the Junior High School.

ARTHUR COTTON
Balboa Postmaster

of the smaller frame cottages were replaced this year with 17 new masonry houses on Ridge and Quarry Roads.

Wartime Balboa

As a town, Balboa was much affected by World War II. A few hours after Pearl Harbor, truckloads of Japanese internees began to stream along Balboa Road, headed for the temporary camp which had gone up overnight at the Quarantine Station on the banks of the Canal. Radio and cable censorship offices were established on Gavilan Road and a three-story frame building hastily put up behind the Balboa Post Office was headquarters for postal censors.

Long rows of tankers lined the Balboa docks and from time to time Balboans glimpsed ships from the Pacific fighting, some of them badly damaged, on their long voyage home. Canteens at docks and at Albrook field hangars were staffed by hundreds of Balboa women. War bond carnivals at Balboa stadium drew crowds

HARRY C. EGOLF
Housing Manager

bushes at them in the early dusk. Another gang of youngsters, now respectable citizens, coiled a large dead boa constrictor in the top of a garbage can and then waited for the collector to make his rounds. They still feel the result was worth the effort.

Balboa Grows

Balboa townsite's first major enlargement came in the late 1920's when 163 houses, including those which now line Amador Road, were built in the area between Plank Street and the sea. Some of these quarters are now in the Naval reservation. The next large-scale growth took place just before World War II when 12-family houses mushroomed in Williamson Place, the Gavilan area, and the section behind the YMCA. The biggest single recent change in Balboa has been the construction of 98 apartments to replace the old frame four-family houses in The Flats. The last of these new quarters was occupied early this year.

Balboa Heights, geographically, is considered part of Balboa, but is still the capital of the Canal Zone as it was designed to be. Its Administration Building dominates the Pacific side landscape just as the two-story, wide-porched house occupied by the Governor dominates its residential section. This big old house had been occupied by the Chairman of the Isthmian Canal Commission in 1907 at Culebra and moved to Balboa Heights in 1914.

Balboa Heights' other big, old official houses, all of which had been moved in from towns "along the line," are gradually being torn down. Some of them and some

ROGER C. HACKETT
Dean of Canal Zone Junior College

of many thousands and sold bonds worth many thousands of dollars.

Second Largest Town

Today Balboa's population of 2,709—or 2,876, if you include Balboa Heights—is exceeded only by that of Paraiso. Its people can attend any one of eight churches, can bank at two banks or borrow or deposit money with the Canal Zone Credit Union, can attend fraternal functions at any one of four handsome lodge halls, or parties at the two USO centers. They can pay their income tax at the Canal Zone headquarters of the Internal Revenue Service or, if veterans, take their problems to the Veterans Administration Office at Balboa

Service Center. Their children can attend school from kindergarten through Junior College, without leaving Balboa.

Balboa is also one of the two centers of commercial activities in the Canal Zone. At the Balboa Terminal Building, 14 shipping agencies handle the business of a great number of steamship lines. Four oil companies, at least one of which was located in Balboa in 1907, have offices and tank farms here. Several shipping and oil companies have residences for their officials in Balboa.

Balboa's New Look

A comparatively new Balboa development, and one which never fails to impress those who have been away for a few years, is the transformation of Balboa Road between the Service Center and Gavilan Road. This stretch is rapidly being lined with handsome commercial, religious and fraternal buildings. The newest of these is a Reading Room for the First Church of Christ Scientist, which is going up next to the Credit Union Building.

This section is particularly attractive at Christmas time when lighted trees, galloping, if unofficial, reindeer, and Nativity scenes bring the holiday spirit to Balboa. Another section which specializes in the Christmas scene is the street off Gavilan Road known as Santa Claus Land.

Balboa, not too long ago, was known as a "boiler makers' town." Today it is, and rightly, the center of the Pacific side community.

EMPLOYEE'S QUARTERS - BALBOA - C.Z.

Tropical Tabogganing

The Panama Canal Review, March 7, 1952, p. 1.

The slope around the Administration Building is one of the most popular of the sledding runs on the Pacific side. The palm fronds carry one, two, or sometimes as many as four customers on a trip. The group above is composed of Mary Smith, Suze Hele, Don Grant, Ellen Anne Rennie and Steve Grant. [See photo on Balboa/Balboa Heights on page 27.]

Summer Recreation Program

The Panama Canal Review, August 1, 1952, p. 9.

In the third Summer Recreation Program, organized for the entire Canal Zone, fathers, pets, dolls, and even mothers' hats, dresses and high heeled shoes played a part in the activities. Fathers and sons pitted skills against each other in horseshoes. "Me and My Dog Day" gave Bowser his place in the fun. Big and little dolls attended a special event at the Balboa Gymnasium with girls all dolled up in their mothers' clothes wearing peek-a-boo veils and evening gowns.

Balboa Flats, Drab But Colorful, To Be Practically Denuded In Next Few Months

The Panama Canal Review, September 4, 1953, p. 16.

Fifty-eight of the 74 four-family frame houses erected soon after the Canal was opened in 1914 are to be demolished this fiscal year. Balboa Flats has been one of the best known residential areas in the Canal Zone; a large percentage of Canal employees have at some time resided in the Flats. If any area could be called drab and colorful in the same breath, it is Balboa Flats. Its drabness comes from the monotony of the typography and the frame houses which are alike as new pennies in shape, size, and color. Much of the color derives from its history and its occupants over a period of almost 40 years.

The history of Balboa Flats—once called in official Canal files "Balboa plain"—dates back to the Canal construction period. The area was once a part of mangrove swamp which made up most of the ground at the foot of Ancon Hill on the Canal Zone side.

Family Quarters for Canal employees at Balboa Heights.

31

New Playground to be Located in Balboa Area

The Panama Canal Review, January 1, 1954, p. 1.

MORGAN AVENUE cottages to be torn down to make room for a new playground.

Construction of a new playground area adjacent to the Balboa Grade School as part of the development of the Flats area in Balboa has been approved by Governor John S. Seybold.

The project will entail the demolition of five cottages and the Junior College Dormitory for boys. It will also require the relocation of a part of Morgan Avenue which runs by the south side of the school building.

The Junior College Dormitory was built in 1921 and was first used as a YWCA. Later it was taken over by the Schools Division and the lower floor has been utilized as a Little Theatre and band practice rooms. The upper floor has been used for a Junior College dormitory since the construction of the Canal Zone Junior College in 1934.

Goethals Memorial Dedication Ceremony

The Panama Canal Review, April 2, 1954, p. 3.

Preceding the dedication ceremony of the Goethals Memorial, located at the foot of the Administration Building steps in Balboa facing the Prado, the Balboa and Cristobal High School joint bands will provide a concert. The memorial monument of "heroic size" was authorized by Congress in August 1935 and is symbolic in its concept. The 56-foot high shaft of Vermont marble represents the Continental Divide and the basins on either side represent the Panama Canal Locks with water pouring from them to join, symbolically, the waters of Gatun Lake with the Atlantic and Pacific Oceans. The shaft rises from a circular reflecting pool 65 feet in diameter. The monument was designed by Shaw, Metz & Dolio, a Chicago architectural and engineering firm, with the Panama architectural firm of Mendez & Sander as associate architects. It was erected by Constructora Martinez, S.A. of Panama, and was completed in August 1953.

WELCOME WAGON service inaugurated

The Panama Canal Review, March 7, 1958, p. 2.

Once notified by the Personnel Bureau that a new family is arriving to the Canal Zone, the Housing Office calls the nearest commissary to assemble a standard assortment of food to provide lunch and breakfast for the next day together with a pair of sheets, a pair of pillowcases, towels, washcloths, light bulbs, soap, garbage-can bags, and other miscellaneous household items.

Temporary "White House" for the Canal Zone

The Panama Canal Review, March 7, 1958, p. 5.

This attractive new house will eventually be occupied by the Lieutenant Governor but right now it serves as the temporary residence for Governor and Mrs. Potter until the remodeling of the Governor's 52-year old house is completed. One striking feature of this house is the wall between the living room and patio which is made entirely of sliding glass panels. The patio/ roofed porch, which over-looks the Canal, is indented to preserve a large tree encircled by an ornamental balustrade.

Great trees shade new official house for Canal Zone's Lieutenant Governor.

Balboa Circle Renamed to Honor Canal Engineer

The Panama Canal Review, September 7, 1962, p. 4.

The Balboa traffic circle opposite the Post Office was chosen as the site of the first Canal Zone monument in honor of a construction era civilian engineer. After undergoing beautification, it is to be officially named Stevens Circle in honor of John F. Stevens, Canal Zone Chief Engineer, who arrived on the Isthmus in 1905 and brought order out of chaos. In the center of Steven's Circle will be a three-sided monument of white Portland cement concrete, with an inscription in Spanish and English in raised anodized aluminum letters. The mahogany trees in the park will be

retained, but planters will be added. A raised center section in the park will be walled with brick.

Stevens Circle is the Town Square

The Panama Canal Review, Winter 1978, p.25.

To a Canal old-timer, it's just the little park in front of the clubhouse. To a Canal history buff, it's a tribute to John F. Stevens. To everyone else, Canal Zone residents and tourist alike, it is a colorful mini bazaar of local arts and crafts. Over the years it has been called the Town Square, the Clubhouse Plaza, The Balboa Park, The Balboa Circle and The Balboa Traffic Circle. The little park has existed from the time the townsite of Balboa came into being as permanent headquarters for the Canal organization in 1912.

It's hard to say just when the artisans of all sorts decided the park was an ideal place to display and sell their wares. First, non-profit organizations held their annual bazaar there. Then came people with plants, home-made items, and even litters of puppies or kittens.

It wasn't long before Stevens Circle was full of Cuna Indian molas, Colombian wall hangings, Costa Rican rocking chairs, Mexican silver, and native birds and monkeys.

In early 1977, Stevens Circle was back on the drawing board—that of a traffic engineer, this time—and the park became neither a true circle nor a square but a circle with three arms.

Celebrated Symbol

The Panama Canal Review, October 1, 1979.

On May 21, 1884, a replica of the Statue of Liberty was formally presented to the American ambassador in Paris by Ferdinand de Lesseps. The idea of a replica of the statue originated with Jack Whitaker, a Kansas City businessman and Scouter of long standing, during the 1951 "Strengthen the Arm of Freedom" crusade of the Boy Scouts of America.

A number of the 7½-foot high copper and bronze statutes were made in a Chicago factory and presented to Boy Scout councils in 39 states. Initially installed in the triangle of land bound by La Boca Road and Balboa Road, it was relocated in May 1972 to the base of the Administration Building, facing the Balboa Fire Station.

Remember the bomb shelters in the early 1940s, like this one in Balboa Heights? *Courtesy of James W. Reece.*

34

"Twilight League" – Baseball

The Panama Canal Review, April 4, 1958, p. 15.

This year's four teams, each limited to 20 players, will be the Old Timers, Working Boys, Balboa High School, and Balboa Boys' Club.

Champs---Baseball on the Isthmus is almost as old as the beginning of the construction period. In 1906, baseball was played everywhere, mostly in the dry season, and players represented various towns or divisions. One of the first things done when Canal headquarters were transferred to Balboa in 1914, was to lay out a baseball diamond between the Balboa corral and the railway tracks.

Mr. Fastlich knew little about baseball but he sponsored the Fastlich Baseball Teenage League in the Canal Zone. He was honor batter during opening ceremonies (shown here).

[Alberto Fastlich was a Jewish refugee from Germany who owned Casa Fastlich Jewelers.]

Above, Commissary Subsistence Team, Champions of Panama Baseball League, Season of 1911-1913. 1. Rutherford, Short Stop; 2. Hodnet, Catcher; 3. Curtis, Second Base; 4. Buchanan, Centerfield; 5. Brievogel, Left Field; 6. McCusty, Catcher (2nd); 7. Meegan, Catcher; 8. Ryan, Right Field; 9. Lacey, 3rd Base; 10. Mosher, 1st Base; 11. Kosher, Pitcher; 12. Sam Carpenter, Mascot. *Courtesy of Lucille Abernathy.*

The Empire Baseball Team, circa 1905-1914. Two of the players are identified as Russell Potter, back left and Fred Whaler, front left. *Courtesy of Dick Cunningham.*

It's Christmas Again

The Panama Canal Review, December 4, 1953, p. 16.

Christmas greetings go to transiting ships through banners like this one at Miraflores Locks. Pedro Miguel Locks control tower will also be decorated with a Christmas greeting banner and lights around the balcony. Gatun Locks personnel are working on their Christmas card, given to all ships which transit the Canal during the Christmas season.

Memories

Santa Claus Lane

Every Christmas morning, Santa and his helper drove up Santa Claus Lane in a jeep decked out with sleigh bells to hand out gifts to all the kids on Oleander Place. When we were growing up on Santa Claus Lane, we didn't have Christmas "gift-opening" at dawn like normal families. Invariably, our radio broke on Christmas morning; so Daddy had to get Ted Henter to drive him and the radio to Mr. Green's to be fixed. We never questioned the necessity of having the radio fixed on Christmas morning, nor did we wonder how it got broken overnight—such gullible children! Meanwhile out on the street, everyone else was riding new bikes or showing off new toys. Syd and I—nothing.

About 9:00, Santa and Ted Henter would careen up the street—jeep horn blaring and sleigh bells jingling. Out of his pack, he would pull a gift for each kid. "Ho, Ho, Ho! Come here, little Frankie, here's a gift for you; and little Cindy and Teddy, here are your gifts. And where's little Rusty Potter?" We kids were impressed that Santa knew all our names.

"Did you like the presents I brought you for Christmas?" Everyone would answer with a resounding "Yes!" Everyone, except us: "We don't know, Santa. We're waiting for our Dad to come home so we can open our presents; he and Uncle Ted had to take the radio to Mr. Green's in Ancon." We really had no clue!

Finally about 10:00, Dad and Ted Henter came home with the radio fixed; and we could have "Christmas." This was the pattern for over a decade—Christmas morning: radio broken, go to Mr. Green's to get it fixed before opening presents.

When we were older, we learned that he played Santa at the Bella Vista Children's Home early Christmas morning. For over twenty-five years, who else but "Red" Townsend with his faithful driver Ted Henter would play a jovial, Spanish-speaking Santa for a houseful of orphans?
- Frank Townsend and
Syd (Townsend) Corbett

Santa Claus Lane (AKA Oleander Place)

For eleven months out of the year, Oleander Place was just a dead-end street in Balboa's Gavilan Area, where kids played kick the can, ring a levio, and hide and seek; rode bikes; and roller skated up and down the street. Ancon (Water Tank) Hill offered a place to build tree forts and pit forts, and to explore through the saw grass for the black panther's lair. But in December, Oleander Place was transformed into Santa Claus Lane, with colored lights, rooftop and yard displays, and decorations rivaling the Las Vegas Strip.

Ted Henter and "Red" Townsend were the instigators behind this transformation. Decorations went up the first weekend in December and came down after the tree burn, the first weekend after Kings' Day.

The Townsends' house had Santa, complete with reindeer and sleigh, and Rudolph perched next to a chimney on the roof. Next door was the Unruhs' snow family in the yard. The Henters' house had a montuno-clad Santa in his cayuco pulled by a red-nosed alligator. At the Potters' end of the street, the Travelers' Palm featured angels at the Nativity.

We Oleander Place kids were blessed indeed to grow up on such a magical street where Santa himself visited and handed out presents on Christmas morning, presents with our names clearly on them, ones that he had forgotten to leave the night before.

- *Frank Townsend & Sydney (Townsend) Corbett*

The Clubhouse

Our main restaurant in Balboa was called "The Clubhouse." There they had a cafeteria, a soda shop, and a bakery where we bought bread that was square, not rounded, and crusty all around. (My sister and I have been known to eat a whole loaf on the way home from a day in the swimming pool.) Also, they had THE best empanadas! They were spicy, flaky and warm. I have since tried many empanada recipes, and this one, I think, is the closest to the Clubhouse's.

Clubhouse Empanadas

¾ pound pork (no fat)
Chop and add to pork:
    1 medium onion,
    1 sweet pepper
    2 cloves of garlic
    1 hard-boiled egg
    10 stuffed olives
    1 large ripe tomato
    1 bay leaf, pinch of oregano
    ½ hot pepper, (optional)
    1 sprig parsley, 1 tbsp. capers
    ¼ cup currants or raisins

Put some oil in frying pan and fry the pork a little. Add the rest of the ingredients and simmer for half an hour. Add salt and pepper to taste. (The empanadas will taste better if this meat filling is made a day ahead.)

Old Balboa Clubhouse

Roll out your dough on a lightly floured board, and cut into rounds. On each round of dough, put ½ teaspoon of the filling. Fold over and flute the edges with a fork. Brush tops with beaten egg. Bake in 400 degree oven until golden brown. Makes 60 small empanadas. You can make larger ones to be part of a regular meal. Enjoy.
- *Edna Hart Crandall*

CLUBHOUSE · BALBOA · C. Z.

New Balboa Clubhouse later known as Balboa Service Center.

Our Own Garden of Eden

Growing up on Mango Street in Balboa we would comb the neighborhood looking for afternoon snacks. We would knock rose apples out of the trees, make mango salad from green mangos and apple cider vinegar, crack open the tamarindo pod to get the sweet seeds and sneak behind the Simonsens' house in hopes of finding the sweet cherry-like fruit we loved so much. Now as adults living in the states (and having to shop at Publix), we look back on that time and realize we truly had our own Garden of Eden.
- Jillian Collins Walker (BHS '84) and Stacy Collins Young (BHS '87), sisters who lived on Mango Street in Balboa, Canal Zone, from the late 1960s to 1977

766 Barnebey Street, Balboa

For 21 years of my first 25, I called 766 Barnebey "home." With the birth of my sister Phyllis in 1927, my dad, Emmett Zemer, met the Panama Canal Government's requirement to have three children in order to apply for larger quarters. The "big" day arrived in 1929 when I was about four. We moved to our new three bedroom, three story, four family quarters, incorporating two upper and two lower dwellings, plus a large full basement with maids' quarters, wash tubs, and parking space for our cars.

We had the lower unit # B, an ideal location with a larger side yard on the corner of Barnebey and Las Cruces. My parents occupied the master bedroom, Phyllis and I the middle one and my brother Bill had the smallest, one corner of which became our mother's sewing room. The only bathroom had a toilet, separate stall shower and a large bathtub. We had a kitchen with an ice box for refrigeration, a gas stove and no air conditioning. Our laundry was hand washed by our wonderful Mabel and hung out on clothes lines in the backyard to dry. We also had a dining room, living room with a piano and covered porch area. My dad used one end of it for his office. I was a teenager before we had a phone or a car.

Living next to us were Beverly and Carol Ruoff, who were the same age as Phyllis and I; we became good friends --played "house" in our basement dressed up in our mother's clothes, jewelry and shoes. Tom, a PC pilot, and Jessie Grimison, lived above us in #A. We moved into their quarters when they retired, and the Lundys became our new neighbors in #B.

Our house was conveniently located. We walked everywhere--to our various schools, church, post office, to the Clubhouse to have cherry cokes, to take in a movie for the cost of 15 cents, and where we three Zemer kids spent the greater part of our youth--at THE BALBOA SWIMMING POOL. Looking back, our life was exciting--we could enjoy the abundance of sunshine, warm temps with balmy evenings and swaying palm trees, complete with gorgeous sunsets, the blue Pacific.

In January 1935, our "baby" sister Shirley arrived. Her crib was put in one corner of her older sisters' bedroom. When our resident owl began to hoot at night, Shirley would cry and ask if she could get into my bed.

I recall one occasion, when our parents had left for the evening, we kids decided to have our own party, led on by brother Bill. "All our friends were doing it, so why couldn't we?" The following morning, we were reprimanded, as the Grimisons had informed our parents what we were up to. We never did it again; we wanted to stay on the good side of Tom Grimison as he used to take us to his Gorgona Beach home in the interior of Panama.

During WWII, our basement became an air raid shelter. I graduated from high school while living there, and upon my return from Boston University, when I went to work for the Personnel Division. Along about 1946 we moved into a cottage on Tavernilla Street, and then to one on Amador Road. In 1950 I relocated to San Francisco. But in my heart, 766 Barnebey Street will always be "home."
- Isabelle Zemer Lively

Crock on the Loose at 762-C Barnebey Street

We lived at 762-C Barnebey Street, Balboa. Our neighbor was building a glider plane and had the crated wing stored in the basement. My cousin and friend were in the biology class at BHS and decided to get a caiman for class and make belts from the hide. They "borrowed" a panga from my dad's machine shop on Dock 19 and took off on the Rio Sucio, which emptied into the Canal near Dock 20. Finding an eight-foot crocodile floating belly-up and thinking it was dead, they hauled it to the dock. It was past midnight, our house was the closest, so they brought it home. They "borrowed" a hand-cart and tied it loosely on the cart. It was full moon, the crock came alive, flipped off the cart and started chasing Bud around the house. The neighborhood awoke in a hurry with all the yelling and screaming. The crock ran under the house and wedged itself between the wall and the crated wing. Albert jumped on its back thinking he'd kill it with a knife to the skull. The policeman on patrol, hearing the commotion, fired a shot that whizzed by Albert's ear, hitting the crock, causing it to flip its tail making toothpicks of the crated wing! Next morning the boys had to get rid of the remains, miss school, and haul it out to the Yacht Club for the tide to take it out to sea. No belt! They barely missed Reform School.
- Bea (Monsanto) Rhyne

Our Special Private Place in Balboa

Living on Las Cruces Street, Balboa, allowed me access to my favorite spot. Directly behind my house and up a slight hill was a small, white transformer building that my brothers and I treated as our personal patio, extended rooftop deck, secret hiding place, whatever it needed to be. For years, I did not climb up on my own. Someone would boost me up so my feet could rest on the large, gray junction box, and then up I went! Sitting on top of that private space was like being on top of the world. At night, it was a dark and creepy place as my brother told his highly exaggerated Tully Vieja and Chicken-Foot-Lady scary stories. During a time when TV was not such an important pastime, it was a place to dream and stargaze. From that little rooftop, on a Friday night, I could see the Balboa High School stadium lights and hear the music and loud cheering. If I looked behind me to Morgan Avenue, I saw the "jungle" as we called it. It was merely the hill to Quarry Heights, but it was thick with underbrush and inhabited by hundreds of gato solos that came down in droves looking for food. The rooftop was also a great place for my brothers to sleep when they were safeguarding our Christmas trees hidden on our back patio roof. When I reached my teens, it became my private beach for sunbathing. Up there with a beach towel, a few friends and a transistor radio, life was good! We shared this special spot with our friends and our duplex neighbors, Carol, Sue and Tommy Coffey. How sweet it was!
- Fran Stabler Meyer

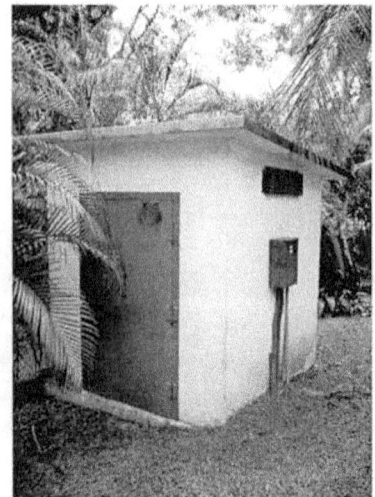

Transformer house. *Courtesy of Fran Stabler Meyer.*

Ernest Horter and "The Taming of the Shrew"

Shakespeare wrote it; Subert Turbyfill produced it; and Ernest Horter turned it into the prize winning "Taming of the Shrew." It was performed in the first high school the town of Balboa ever had. In the center of this school was a wide-open patio, a larger square grass plot where trees and bushes grew and rain entered in the rainy season. Squeezed into the grass plot were 400 steel folding chairs.

At one end of the patio, Director Turbyfill had an elevated stage constructed, six feet wide and twelve feet long. In this patio, in 1941, Subert Turbyfill presented the best and finest show he ever produced. It must have been the best—he put it on the cover of his book, "My Panama Canal Theatre Adventure," published in 1949.

To the left stood Ernest Horter, the haughty and demanding Petruchio, reigning over all. To the right Jeanne Flynn, the fiery Kate and supposedly Petruchio's bride. Miss Flynn "lived her part through every minute of action," so said the PANAMA AMERICAN, Panama's English language newspaper.

Some others in the play were Alan Bentz, Fred Brugge, George Muller, Marilyn Sanborn, Eleanor Sullivan, many others. Miss Eloise Monroe, a new teacher at Balboa High, with a staff of 15, produced the elaborate costumes.

1941 was the last year this building was a high school. The following year it became a much-needed elementary school. A new building over near Roosevelt Boulevard became the new high school.

Turbyfill sold out and surprisingly required a second printing due to unexpected demand throughout the United States. Ernest Horter became a LT (jg) in the US Navy and served during World War II, 1943 to 1946. Recalled to active duty again, alas, Ernie went on to die in a plane crash on his way to Korea. Jeanne Flynn is still around, doing fine, and longing for the dear ole Canal Zone.

- Jeanne Flynn Stough

WW2 Memories in Balboa

I was almost four but remember vividly the first day of the war. I lived in a four family house at the end of Carr Street. The Olivers and the Gaus also lived in this building. What I remember the most is all the excitement and talk about the idea that the Japanese were going to attack the Canal. Workers were sent out to paint the top part of the street lights black. The next day the building of bomb shelters was started. One went up right beside our house. It became a playground for us kids. Someone gave us little guys a disabled 30 caliber machine gun on a tripod. It took three or four of us to carry it around. Every kid in the neighborhood was playing war games at this time, and we were the only ones with a real gun. Of course, it got stolen.

The most deep seated memory that I have is of one of the older of the four Bobletts boys. He somehow got his hands on a hand grenade. His younger brother and I were watching him try to take it apart in the bathroom of their house. For some reason we two little guys walked out of the bathroom into the living room before it exploded; neither of us was hurt. The older boy spent the next six months in the hospital.

My mother became a Red Cross volunteer. Her assignment was to serve coffee and doughnuts to the troops that docked at Pier 18 in Balboa prior to transiting the Canal. Often she would take me with her, which I thought was great because most of the time the sailors on the ships were on their way back from the Pacific war zone and many of them would give me a sailor hat.

Another war memory I have while living in Balboa is of the folks' using mango tree branches for Christmas trees.

- Jim Des Londes

Four Youths Are Fined For Sosa Hill Fire

The headlines of the Panama American article announced "Four Youths Are Fined For Sosa Hill Fire." As it happens, I was one of the "youths" who had been arrested for participating in a long-standing and much-hallowed Pacific side tradition--setting fire to Sosa Hill.

At the end of each dry season, the Fire Department would schedule a controlled burn of this Balboa landmark in order to reduce the fire hazard that the hill presented in its tinder-dry state. Our goal was to beat the fire department to the punch.

Thus, in 1955 at the close of the dry season, a group of four boys, Jim Fitzpatrick, Jimmy Boughner, Norm Dixon and I, set out to burn the hill--on the very night that most members of the CZ Fire Department were attending the Annual Firemen's Ball at the Hotel El Panama. We were really looking forward to watching the firefighters respond to the fire from the dance, dressed in their formal attire.

In the dark of night we ascended Sosa Hill from a point behind the Balboa swimming pool. Although we were totally successful in setting the hill on fire, our escape plan left much to be desired. After setting the fire, we descended towards Tavernilla Street, where our route took us directly under the house occupied by the Chief of Police and his family. We didn't notice the chief, and ran right by him. Unfortunately, the chief did notice us; and, as was quite common in the small-town world of the Canal Zone, he knew each of us by name.

The rest is history.
- Paul D. Glassbum

Sosa Hill rises behind Balboa Theater.

Platform on Balboa Hill

The Canal Record, May 13, 1908, p. 293.

A platform has been built on top of Balboa Hill, about three and one-half miles from Gorgona, by Mr. Fletcher Stevens, from which both the Atlantic and Pacific oceans are visible on a clear day without the aid of a glass. The line of the Canal can also be seen for a long distance as can likewise most of the towns in the Zone. An American flag, which has been raised above the platform, is visible from Gorgona.

Among the several hundred photographs taken by Mr. C. L. Chester during his visit on the Isthmus are two which were taken from the platform on top of Balboa Hill, near Gorgona, on June 2. In one of the exposures the camera was facing the Atlantic Ocean and the other the Pacific, and in each photograph the ocean is distinctly shown.

Empire and Balboa:
R.B. Potter

Did you ever know that there were costume parties at the Pacific Sailfish Club, what we later knew as the Balboa Yacht Club? This one took place in the 1920's, and my grandparents, Russell and Linda Potter, were there. He is the clown in the top row, and she is the veiled lady in the bottom row. This was 20 years after Russell went to the Canal Zone in 1905.

When he arrived, men were lined up on the dock waiting for the next ship out. Perhaps the long hours and hard work discouraged them. Or the heat and bugs and the disease. Or the endless monotony of the jungle. But all this didn't discourage Russell. He figured he'd give it a try, stay for six or eight months and then go back home; he'd gone down for the pay. And the pay was good, $100 a month, enough to get him to quit his job with the Pennsylvania Railroad and go down.

Halloween party at the Balboa Clubhouse. *Courtesy of Dick Cunningham.*

When he got off the boat, he was 21 and they took all of the new employees to the canal headquarters in Cristobal and gave them mosquito netting, a tin pitcher, and tin pail. Then they were put on a train to a construction camp called Empire and were shown their quarters. There were no walls and all you got was a cot. They were told, "Here you are; you're on your own." When he arrived in Empire, men were still dying, coffins were stacked at the edge of camp and workers in the mess hall were gulping quinine and rum.

Yellow fever was still going around. Men in Quarters would suddenly get fevered and the next minute you'd hear they were in the hospital; some came back, and some didn't. They had to drink quinine all the time for malaria. It was awful tasting stuff. They tried to get the men to drink it by mixing it with wine. When that didn't work, they would mix in a bit of rum. They couldn't keep the pitcher coming fast enough. Wonder why?

In hospitals, under the legs of the beds, were pans of water to keep out the ants. Nobody could understand how a man going to the hospital with typhoid would end up dying of malaria. Then they discovered that yellow fever and malaria were spread by mosquitoes. And those pans of water under the patients beds were where the mosquitoes bred.

There were many men and women who came to the canal in the early years and spent most of their lives until retirement. Russell started as a stenographer and ended up being assistant to the chief engineer, working out of the Admin building in Balboa Heights He and my grandmother, Linda, lived on Morgan Avenue in Balboa in a single family home until 1947, when he retired. All these workers were pioneers in a masterful effort to conquer nature. The Panama Canal was their life.
- *Dick Cunningham, Grandson*

1947, Brownies are parading on the Prado. Notice the envious children watching from the curbside. The birthday party photo op, traditionally held on the steps on our apartment in the Flats.
- *Courtesy of Edna Hart Crandall*

Balboa Swimming Pool, 1927.

Balboa Yacht Club.

The Working Boys

Left, In 1947, this was the Champion Football Team of the Canal Zone - The Working Boys. Recognize any of them? Left to right, front row: C. Fisher, unidentified player, Carl Newhard, Earl Daily, H. Townsend, Stanley Whaler, Middle row: B. Lockridge, Ross Cunningham, H. Clarke, F. Curtis, T. Malone, J. Hickman; Back row: J. Brown, J. DesLondes, D. Leonard, G. Tarflinger, and F. Wertz.

My father, Ross Cunningham, was one of them, and many of these men were close friends of his and our family. I attended school with many of their chidren. Ross came to the canal in 1928, and his first job was counting banana bunches being taken off the banana boats. He was actually on the Silver payroll but very soon he was hired to work in the Accounting Department in the Administration Building.

When I was in grade school at Balboa Elementary, I visited his office in the basement Vault where all the money and Commissary coupon books were stored. Once he said, "If you can carry that bag of silver dollars out, you can have it." Well, I could not even lift it off the floor, so I didn't get rich.

Ross also played softball for the Admin team. It was fun to watch. They usually had a keg of beer on hand to quench their thirst. He was also a square dancer and called the dances at the Ancon Playshed for many years.
- Dick Cunningham

The Elks Club

When I was in junior high school, we used to visit the Elks Club in Balboa on La Boca Road. There we could play pool, ping-pong and other games. They also sold a great lunch although we couldn't get beer.
- Dick Cunningham

BALBOA - CANAL ZONE

My Canal Zone Home in Balboa

A home is a place of residence or refuge. "Home" is also used to refer to the geographical area (whether it be a suburb, town, city or country) in which a person grew up or feels they belong. The Canal Zone was my home. Growing up we moved throughout the Canal Zone so I cannot claim one townsite as home. Instead I want to share the homes we lived in and certain memories from them.

In January of 1967 my dad, Paul Hurst, got a job with the Panama Canal Company, which allowed us the privilege of living in the Canal Zone. I was four and my brother Paul was five. We lived on Tavernilla Street in Balboa. My mother Delia tells us that we were so happy. She tells us that every time family came to visit we would give them a tour of the house, including the bathtub! It was a four family house. From there we moved to a house right next to the post office in Balboa, but we only lived there about six months. My brother Harris was born while we lived in Balboa.

In 1970, my dad was promoted and we moved again, this time to Diablo. For some reason my parents did not like it there and within three months we moved to Balboa again. We lived one house over from where we had lived before, a four family house right down from the post office. This is where I met Linda Magee who became my lifelong friend. I remember walking together to and from Balboa Elementary School, stopping on our way home to buy popsicles and rock candy from a family who sold them.

Linda had moved to Los Rios and I wanted so much to move there so we could spend more time together. The Housing Division would post the houses available at the Balboa Post Office. You were chosen by your years of service and your grade. Every chance I got I would look for a house in Los Rios. Finally one came up and I convinced my dad to put in for it, and we got it. In 1977 we moved to Los Rios and my parents lived there until my dad retired from Marine Traffic Control in 1991.
- Helen Hurst Loera

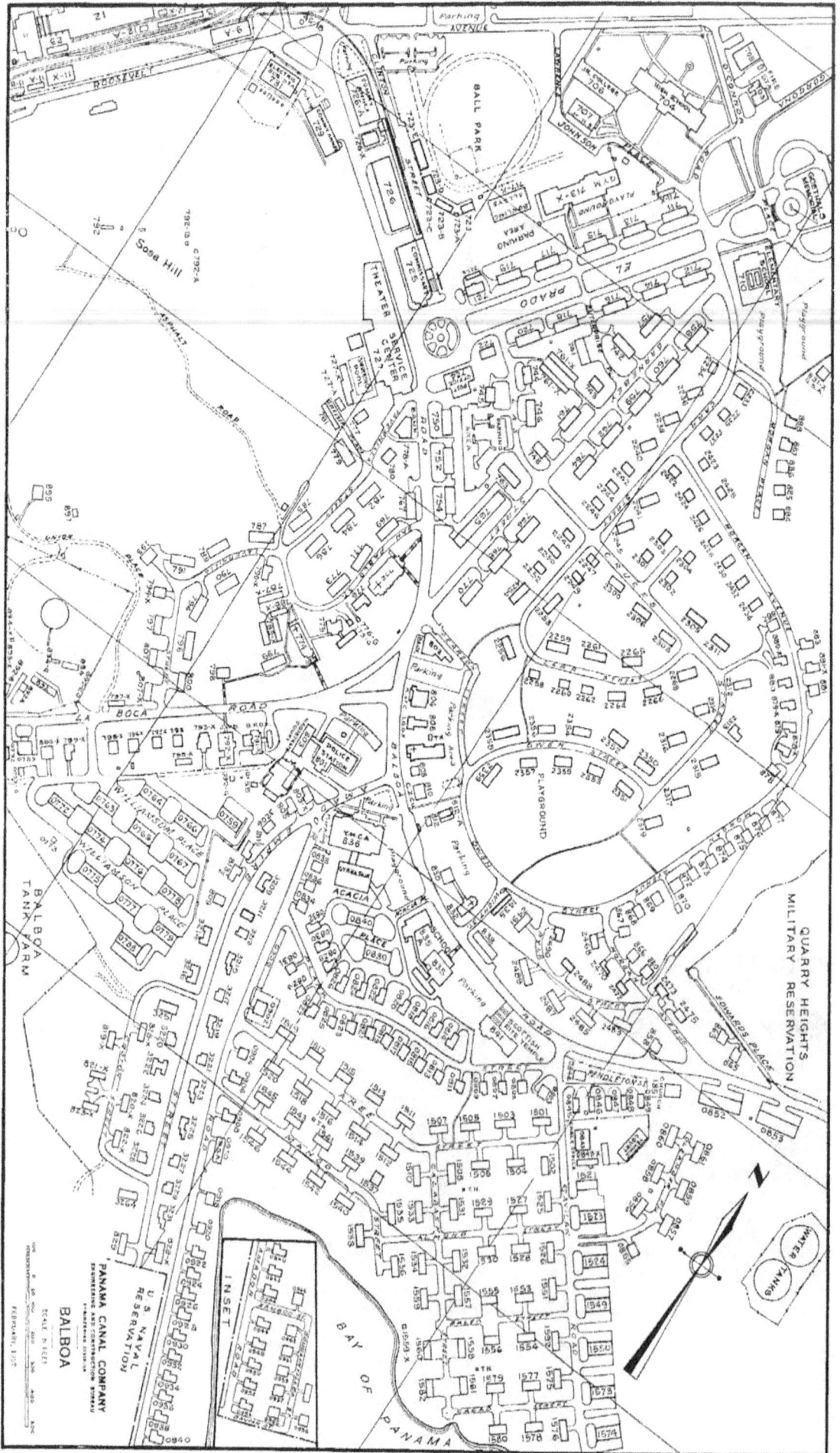

Cardenas

Sponsors

Joan McCullough Ohman
Bob & Cheryl Russell

Cardenas

The Panama Canal Review, June 6, 1952, p. 4.

An entire new town site is being developed north of Corozal for local rate employees on the Pacific side. The contract for the project was awarded to Macco-Panpacific, Inc.

Photo shows progress of Quarters Program.

Cardenas Site Work Over One-Third Done

The Panama Canal Review, August 1, 1952.

Approximately 1,000,000 cubic yards of earth are being moved in the major clearing and grading job required in the preparation of the new local-rate town site of Cardenas. The work is being done at a cost on $1,225,000. It is the second largest project of the 1952 housing program on the Pacific side.

The contract was awarded last February in two parts, one for the clearing and grading of approximately 175 acres of hilly and heavily wooded land north of Corozal; and another for the installation of an access road from Galliard Highway. The new highway will be reinforced concrete. It will be built from Galliard Highway to the new town site. The new road leaves the main highway a short distance from the existing entrance to Corozal Hospital. In addition to the highway, a large storm drainage structure is being installed as well as a large sanitary sewer, 4,230 feet in length, which will connect with the Cardenas River north of Gailliard Highway. Two 250,000-gallon capacity water tanks are being erected at the new town site. The contract completion date for the overall project is next May.

Housewarming Time for FAA

The Panama Canal Review, March 3, 1961.

Nestled snugly against the gently sloping face of a hill leading down to the Cardenas River and located between Fort Clayton and Corozal is a new Canal Zone community of 90 housing units all of which are to be occupied by employees of the Federal Aviation Agency in the Zone and their dependents. The 133 employees of the FAA in the Zone operate an International Flight Service Station from the Civil Affairs Building in Ancon, thus fulfilling an obligation assumed by the United States in the "Aviation Agreement Between the United States and the Republic of Panama" in 1949.

The new community taking its name of Cardenas Village from the river which flows past the base of the hill on which it rests, will be officially opened on March 12, when Gov. W.A. Carter will officiate at a ribbon-cutting ceremony which will mark the beginning of a public open house at several of the new dwelling units.

Most of the employees of the FAA and their dependents have been living in the former Navy Community of Rousseau on the west side of the canal. Other FAA personnel now live in the Cocoli, Ancon, Balboa and the Republic of Panama. The housing units in Rousseau which are to be vacated by FAA personnel will be returned to the Navy after the present occupants have moved out.

In addition to the 90 housing units ready for occupancy at Cardenas Village, 30 more units are scheduled fro completion during the next 5 to 6 months. Construction of the housing units by H.A. Lot, Inc. of Houston, Texas and Drake of Panama have been done with the Canal organization acting as contracting agent. The Canal organization will also provide the new community with police and fire protection, water, electricity, sewage disposal, garbage collection, and similar services.

Left, Type 501, were 628.8 sq. ft. apartments. Right, were two family units featuring 1,038 sq. ft. of living area with two bedrooms and one and a half baths. *Courtesy of Ed Ohman.*

Cocoli

Sponsors

Carl N. Berg
Charles W. "Chuck" Hummer
Joan McCullough Ohman
The Walkers: George & Mayno, Mickey, Fred, Jeanne & Carole

From *Cocoli Verses: Cocoli, Canal Zone*

by Carl N. Berg

Aerial view of Cocoli.

Above, Four-family house at 2386 Durante Place in The Hollow. Below, Two story duplexes like this at 2532 Nicobar Avenue were considered highly desirable by Canal Zone families. Taken in 1993. *Courtesy of Carl Berg.*

During the period of American canal construction, 1904-1914, and the decades that followed, town sites were established by the U.S. Government on both ocean ends of the Canal Zone. Cocoli is located on the west bank of the Panama Canal on the Pacific Ocean side between the Rousseau housing area and the Miraflores Locks. Bruja Highway is the main roadway that passes through Cocoli. It begins at Thatcher Highway, near the front gate at Rodman Naval Station, and ends at the west bank side of Miraflores Bridge.

In 1939, the U.S. Congress authorized the construction of a third set of locks. Cocoli was one of three town sites started to house and support the work force for the completion of the Third Locks Project.

Cocoli had a clubhouse with a cafeteria, billiard parlor, movie theater, barber shop, shoe shop, beauty parlor, and a bowling alley. Bingo and dance lessons were a few activities provided at the clubhouse. A commissary served as the town grocery and general store. Cocoli also had a post office, police station, and fire station. An elementary school, gymnasium, Little League ball park, several playgrounds, and public works structures made up the rest of the facilities in Cocoli.

There were a variety of residences or quarters, namely one-story single family cottages, one- and two-story duplexes, two-story four-family houses, and two-story multi-unit bachelor quarters, all having carports at street level.

Cocoli Elementary School was set back from the curb of a small one way traffic circle connected to a two lane black top driveway off Tamarindo Avenue.

There were three churches in Cocoli: St. Andrews Episcopal Church on Second Street, Saint Teresa Catholic Church on Bruja Road, and the Baptist Church at the base of Cocoli Hill on Bruja Road.

Driving through Cocoli toward the Miraflores Bridge, or north, everything to the left side of the highway can be called the "hill" side; everything to the right can be called the "town" side. The few structures on the hill side were: one residential fourplex, the Chinese Garden, the gun club, the police station, the Baptist Church, the Scout shack, the Roman Catholic Church, a small maintenance building or two, and the Local Rate housing area.

Everything else, over ninety percent of the streets, buildings, and common areas, was on the town side of the highway.

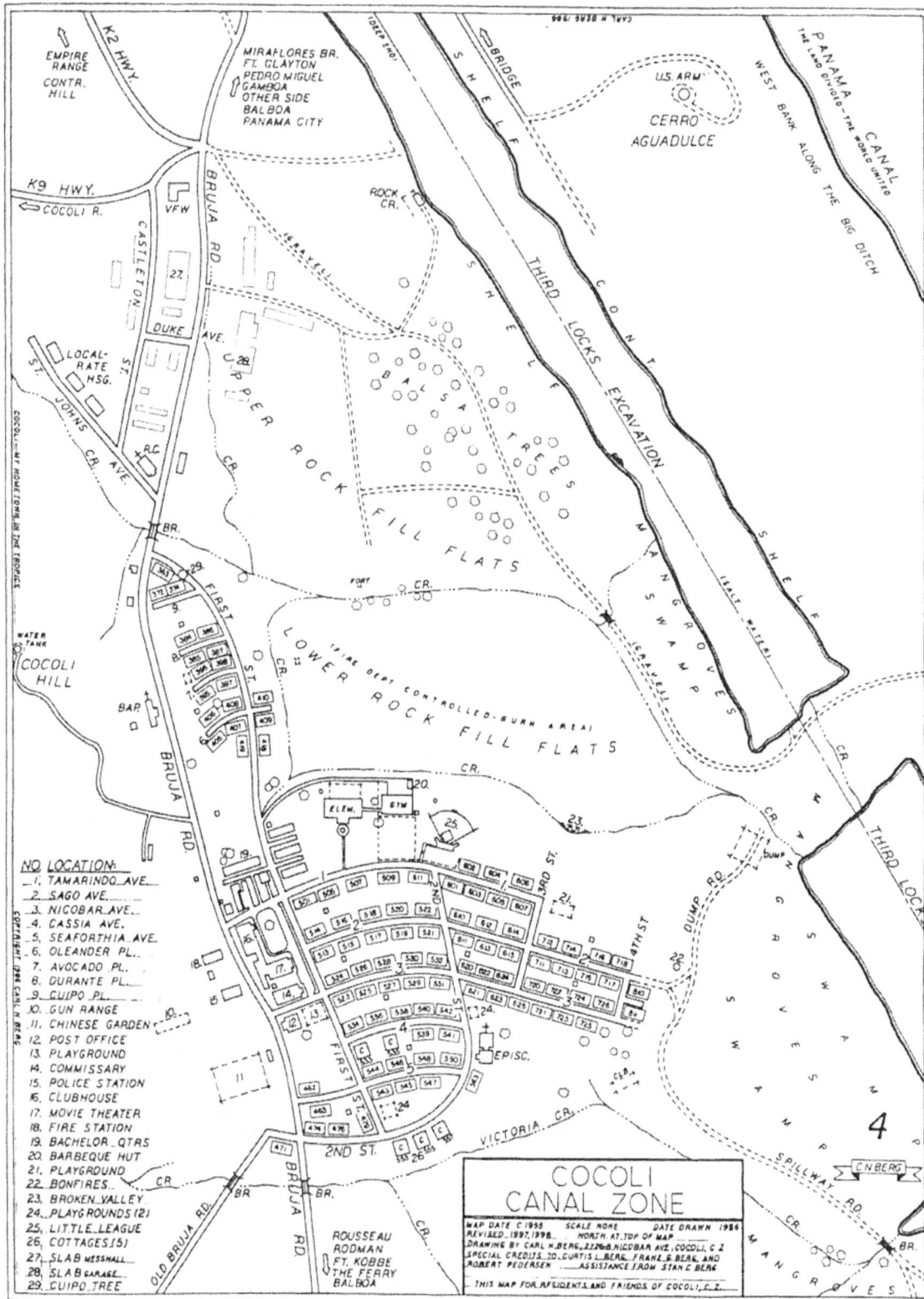

NO LOCATION:
1. TAMARINDO AVE.
2. SAGO AVE.
3. NICOBAR AVE.
4. CASSIA AVE.
5. SEAFORTHIA AVE.
6. OLEANDER PL.
7. AVOCADO PL.
8. DURANTE PL.
9. CUIPO PL.
10. GUN RANGE
11. CHINESE GARDEN
12. POST OFFICE
13. PLAYGROUND
14. COMMISSARY
15. POLICE STATION
16. CLUBHOUSE
17. MOVIE THEATER
18. FIRE STATION
19. BACHELOR QTRS
20. BARBEQUE HUT
21. PLAYGROUND
22. BONFIRES
23. BROKEN VALLEY
24. PLAYGROUNDS (2)
25. LITTLE LEAGUE
26. COTTAGES (5)
27. SLAB MESSHALL
28. SLAB GARAGE
29. CUIPO TREE

COCOLI
CANAL ZONE

MAP DATE C 1955   SCALE NONE   DATE DRAWN 1986
REVISED 1997, 1998.   NORTH AT TOP OF MAP
DRAWING BY CARL H BERG, 2326A NICOBAR AVE, COCOLI, C Z
SPECIAL CREDITS TO CURTIS L BERG, FRANK E BERG AND
ROBERT PEDERSEN   ASSISTANCE FROM STAN C BERG
THIS MAP FOR RESIDENTS AND FRIENDS OF COCOLI, C.Z.

The new Bruja Road replaced the old Bruja Road, a long-abandoned, mostly forgotten (except as an occasional lover's lane) narrow blacktop road, overgrown high on both sides with jungle. The old road connected Cocoli to Rousseau and to NAD. In Spanish the word *Bruja* means "witch." *Your car will get jungle scratched if you drive in there.*

The Chinese Garden was a small, corrugated roof-covered building facing Bruja Road. It was open at the front and both sides, where tables of fresh fruit and vegetables were displayed for sale. The gardener's enclosed area at the rear, was a tiny kitchen, sleeper, and store room. *The large papayas, sugarcane, and other fruits and vegetables look especially good.*

A rain (storm) ditch in the Hollow begins below street level at the large circular concrete opening on the school side of the road bank, at the intersection of First Street and the gymnasium access road. It ends past the Hollow, where it connects to the small creek passing under the bridge near Saint Teresa's Catholic Church. The kids living in the Hollow take the grass footpath that crosses the rain ditch. An effortless jump gets you to the other side. It is the back way to get to and from Cocoli

School. At times, the ditch has little or no water. But during the rainy season it often swells to several feet deep and maybe eight to ten feet wide, with a very swift current of water. This rain swollen ditch becomes a brown water playground. The kids allow the current to float them quickly several hundred feet downstream, about as far as Avocado Place. Another fun activity is to have small boat races. A short piece of a 2x4 board makes a good boat. An old inner tube makes the best personal float. *Make sure you wear long pants, because there are underwater obstacles you might not expect.*

Famous trees of Cocoli were: the

47

Corotu Tree, a very large tree whose branches tower over nearby homes and lower limbs stretch out over the road. A small "fruit," when ripe, has a dark brown, wrinkled, sticky, pithy outer shell, or pod, resembling the human ear. Inside are hard brown beans about the size of peanuts. They are not edible. Kids rub the beans on the sidewalk to heat them and press them to another's skin, thus the nickname "burnie beans"; The **Jobo Tree,** has a sweet, red-yellow-orange, fruit with a thin skin. It is edible, tasty, and juicy. At times it can be bitter and grows to the size of a cherry tomato; the **CuipoTree,** is found in the middle of 1st Street. The tree stands about 60 feet tall and about 50 inches in diameter. This barrel-trunked giant was left standing for the appreciation of the Zonian. Both lanes of traffic go around the tree; **Tamarind trees** are planted along the south side of Tamarindo Avenue. This tree is probably native to India. In Spanish, tamarind becomes Tamarindo. The fruit is best consumed when ripe, in its brown, sticky, sweet bean pod or made into juice.

The first housing authority in Cocoli was the Panama Canal Commission. The U. S. Navy gained control in the early 1950's and, finally the U.S. Army. At times some of the civilian-military, or military- military authority may have overlapped.

Zone Youngsters Learn Skill and Safety Handling Guns at Junior Rifle Club

The Panama Canal Review, June 5, 1953, p. 9.

In the 1950's, youngsters of Cocoli learned the skills and safe handling of guns at the Junior Gun Club. George M. Sylvester, manager of the Gun Club, which was affiliated with the National Rifle Association, was responsible for teaching the young boys, ages from 9-18. The type of gun, a .22 caliber Remington 521T (target rifle) especially made for youngsters, was used by the boys. The first time a boy shot a good set of targets, the target went home with him so that his family could share his excitement of accomplishment. Boys never went home empty handed from the Club prize shoots. Rather than medals, the boys received things they liked such as jack knives and model cars. The rifle ranks began with pro-marksman, marksman first class, sharpshooter, expert and finally distinguished rifleman. In the early 1950's, 15 year-old James A. Hale, Vice President of the club, was classified as "expert rifleman" with six others fast approaching that mark. Gerald Hendrickson, at age 17, was the only Distinguished Rifleman, the only such achievement in the history of the Club up to that time.

Members of the Cocoli Gun Club are shown here with their manager, George M. Sylvester, right, at one of the regular Saturday morning practice sessions.

Cottage at 2553 Second Street, was converted to the Community Center by the U.S. Army. *Courtesy of Carl Berg.*

Cocoli Clubhouse.

CLUBHOUSE - COCOLI - CANAL ZONE

Memories

Cocoli . . . Where the Living Was Easy

Nestled on the west bank of the Panama Canal, just past U.S. Naval Station Rodman off Bruja Road, lies the enchanting town site of Cocoli.

Cocoli for many years was my home. In its heyday it boasted everything from a clubhouse to a clinic dispensary for common prescriptions. There were four family houses, duplexes, cottages, single apartments for the bachelors, and also 12 family units. It was a bustling town like those carved out of the jungle of the Panama Canal construction days. Yes, living in Cocoli was fun! The main part of the town was on a mesa surrounded by a series of low hills to the west that led toward K-9 Road and Miraflores Locks. A portion of the town dropped away rather sharply to the west past the Cocoli Elementary School. It was affectionately known as "Sleepy Hollow." Yes, just like the "Halloween Tale" we all had to learn in grade school. So on a spooky night when the moon played hide-seek among the low dark clouds, you just might see the headless horseman come out of "Sleepy Hollow." We little kids would run all the way home! It was really scary attending first and second grades at Cocoli Elementary.

Then we graduated and went on to junior high in Balboa. It was all a new experience with the bus and all, but we survived to hit the big time by starting the experience of a lifetime at Balboa High.

- *Louis Joseph Barbier, Jr.*

A Mango Tree Grows in Cocoli

When I was a kid growing up in Cocoli, I lived in a wooden duplex on Tamarind Avenue. In my back yard there was a mango tree, which I called, Sammy Ho. It was a very beautiful tree that produced some of the best mangos in the world! It had large green thick leaves that waved at me each morning. I could see it from my bedroom window. It was a marvelous tree that had been married to a couple of tamarind trees. Don't laugh. This is a must! Mangos are the kings of the tropical fruit! The founders of Cocoli knew that for outstanding mangoes they had to be married to tamarind trees. The tamarind fruit is the queen of the tropical fruit! The tamarind tree tends to be a bit tart and sassy . . . just what a good mango-producing tree needs. We had plenty of tamarind trees in Cocoli.

Both the mango and the tamarind trees are great for climbing. They will grow to 50 feet. Wow, what a bird's eye view of the world you can see from up there. Sammy Ho was maybe 30 feet tall and still growing. I had a favorite spot I would climb which was level with the tin roof of our tropical quarters.

Sammy Ho produced large mangos with red shoulders. Teddy Roosevelt would have said, ". . .These mangos are bullish, Cocoli; get me a sack full for eating later in my room at the Tivoli Guest House."

- *Louis Joseph Barbier, Jr.*

Battleball!

In my elementary school years, especially grades 4 through 6, Cocoli School kids played battleball without let up not only after school but during weekends and summer vacations. For us it was a passion, an obsession, and we couldn't have lived without it. Our gymnasium, on the upper of two floors, was connected to the school by a 100-ft.-long, open but roofed walkway. Just inside the double doors lobby entrance was a well-placed jukebox. At the end of the lobby there was a coach's office, players' lockers, showers and restrooms for boys and for girls. On the playing side of this front section was the basketball court, maybe smaller than regular size, with bleachers space at both sides and a small elevated scorer's box for two.

A dodge ball variation. Two teams (no player limit) shoot it out with basketballs, volleyballs, and dodge balls, each team keeping to its side of the court. A dozen balls are placed on the center line, and then each team rushes for the balls. Players hit directly must retire to the side. Often a good player is targeted by many opponents' balls. A thrown ball can be deflected by one being held; floor-bounced ball hits don't count. The last player on a side can cross the center line, but he can't be chased across it. Last man standing wins for his team. Front team: Timmy, Carl, Mickey, Nancy, Johnny, Minor, Mary, Georgie. Back team: Stanley, Shorty, Frankie, Jeannie, Warren, Judy, Kathy, Billy. GO!

- *Carl N. Berg*

Cocoli School, 1977. *Courtesy of Carl Berg.*

Finders Keepers

In 1957 we moved from 2398-C Avocado Place in The Hollow to 2726-B Nicobar Avenue in the main housing area. Before that, however, the three Berg brothers were no strangers at the Cocoli trash dump. The messy open site, roughly 60-feet by 80-feet, was at the east end of town, about a hundred-fifty yards from the last house on Sago Avenue. The residential blacktop pavement ended at that house, then continued in gravel a short distance before turning toward the dump, not far from the south end of the Third Locks Cut. The dump was maybe one-eighth mile away, about a five minute walk from our duplex on Nicobar. A narrow dirt trail connected the dump to the salt water at the Cut.

We were curious and unashamed. Not only did we occasionally look in neighbors' garbage cans for salvageable items, the dump at the end of town was a much better place to shop. An abandoned, tires and wheels-missing late-1940s black Cadillac was off to one side; it had no business being left there. Tossing common trash aside, we focused on objects that might contain a treasure or two: shoe boxes, suitcases, cardboard cartons, unopened bottles, cigar boxes, cigarette packs, jewelry cases, tied bags, clothing, tool boxes, and such. Special finds were: Playboy and other men's magazines, cigarettes, comic books, wrenches, radios, playing cards, fishing gear, military equipment, school supplies, auto parts, and such. "Hey, look what I found!" meant something worth keeping might have been found.

- Carl N. Berg

Keep on Walking

Cocoli's flat, lower plateau was a dump area of Third Locks Cut rocks spoil, located east of The Hollow housing area and north of the main housing area. Three connecting creeks (see map) defined the plateau's perimeter. Annually, during late dry season, the entire plateau was burned off under fire department control. In March 1951, the three Berg brothers, Carl (9), Stanley (7), and Franz (5) had a small wood fort about 100 feet in from the start of the high grass nearest First Street. To properly play Indian there, we would set aside our regular clothes and put on loincloths and face paint. With cane spears in hand we would do a war dance around a small fire whose flames went no further than a three foot circle of rocks. We still don't know exactly how the flame transferred to the nearby high grass. But when we realized what was happening we panicked and tried to beat out the accidental fire by throwing boards on it, causing it to spread quickly beyond our control. We grabbed our clothes and made a straight run for the area behind the Little League outfield fence. Soon fire trucks were trying to control the large plateau fire. We sneaked up to Tamarindo Avenue and slowly found our way back down to The Hollow. As we approached our house on Avocado Place, Dad was out front talking to a neighbor. When we passed by, he simply said, knowingly, "Keep on walking."

- Carl N. Berg

Cassia Avenue.
Courtesy of Carl Berg.

Episcopal Church Chapel, 1993.
Courtesy of Carl Berg.

Cocoli Townsite, circa 1945

When World War II commenced on December 7, 1941, I lived in Gatun. During most of the wartime I lived in the new town of Cocoli on the West Bank, Pacific Side, in the Panama Canal Zone. The town was built in the early 40's and was similar to Margarita on the Atlantic Side.

I attended Cocoli Elementary School through grade 6. We had a town with a clubhouse, commissary, gym, and a baseball park. It was a great place to be during WWII and our names were Bright, Godby, Gorman, Halsall, Maphis, Ruoff, Schmidt, and Tinnin and many others.

During the war years, toys were made of wood or paper. No bikes or metal toys were available, as the metal went to the war effort. Most of the military was at Fort Kobbe, Howard Air Force Base (AFB), and Rodman. At that time, the Cocoli kids would bus to Howard AFB and ask the airmen for stuff. The generous airmen would give us a life raft to row on Cocoli River and Miraflores Lake. They also gave us life vests to swim with. We asked for and received flare guns and flares to celebrate the 4th of July. All standard fireworks were not to be had as they went to the war effort. We were even allowed to play the slot machines in the Officers' Club. The Rodman Navy gave us baseball equipment to use on our baseball field. Other towns wondered how we got all the good stuff. We said it was a military secret.

Times were tough and it was difficult to get supplies and normal goods from the US during the war; however—for the "Kids of Cocoli," it was a fun time, and we felt fortunate to be able to obtain equipment for our use and enjoyment.

At the time of the cease-fire with Japan, on August 15, 1945, the town of Cocoli celebrated by going outside and honking the car horns. The town of Cocoli would gradually phase out in the 1950's, but the memories are still with the "Kids" of Cocoli.

- Jerry Halsall, a Cocoli Kid.

Coco Solito

Sponsors

Gerry DeTore
Mary Elizabeth (Beth) Bialkowski Lozano

Random Memories of a Five Year Old Living in Coco Solito

My dad, Wendell Cotton; my mom, Lois Cotton; my brother, George, circa 1945

It was fun living in Coco Solito on the Atlantic side after my dad (Wendell Cotton) was transferred from Rodman in the mid 1940's. The Humphreys lived in the next building over. Mr. Humphrey was a fireman, as I recall, and the fire station was right across the street from their building. My two-year-old brother, George, used to get up about 5 o'clock in the morning and go across the street to visit the firemen. They would "entertain" him until my mom was up, and then they would bring him home.

My aunt and uncle, Frances and Dan Hennessey, and my cousins, Mary and Carolyn, lived next door. Our parents decided that it was "too much trouble" to go down our stairs and up their stairs to visit each other so they had a door cut out between the two apartments. I don't have a clue how this was accomplished, but my dad knew all kinds of "magic tricks" and could get things done — especially when it came to housing! So now we had two kitchens. We had ducks and turtles downstairs under the house. When cats discovered them and began to eat them, we moved them upstairs into one of the kitchens. I can't imagine the stink, but luckily our parents were in their 20's and had not yet become "adults."

Across the stairwell from Frances and Dan lived another aunt and uncle, Snookie and Mac McCullough. They had two kids, my cousins Judi and Tommy.

All of our dads were in the Navy so you can imagine the parties! We kids loved it.

At Christmas, we kept all the old trees in the stairwell upstairs from us! Since no one was living up there, it seemed a perfect place as no one could "steal" them before the Christmas tree burn! It is a wonder the place didn't burn down with all those dried out Christmas trees stashed upstairs!!!

The blimps used to fly over Coco Solito and drop leaflets for savings bonds. We had a great time playing "store" using those leaflets for "money." It became a contest to see who could gather the most leaflets.

Mr. Green was the manager of the Coco Solito Commissary and later the manager of Cristobal Commissary.. He was the same Mr. Green who could snap a commissary book out to the exact amount with a flick of his wrist! He knew everyone by name! We kids would go to the commissary and "help ourselves" to kool-aid packages, candy bars, and what-have-you, and Mr. Green never said a word. However, when our parents would show up at the commy, Mr. Green would present them with a "bill" for our "purchases."

I was constantly told that I should not swing on the rope clothes line under our house, but of course I did. One day the rope broke and I woke up at the Colon Hospital emergency room. Luckily, there was no serious harm done! Needless to say, I did not do that again!

One day my cousin Judi and I missed the bus home from Margarita School. I had convinced Judi that she should accompany me to the bathroom. She was in first grade, and I was in kindergarten. We decided we had no other choice but to walk home to Coco Solito. It never occurred to us that we could go to the Principal's office and use the phone. We knew we were in big trouble! So we started off down Snake Road. We were scared to death!!! We expected snakes, alligators, lions, tigers and other wild beasts to jump out of the jungle and eat us alive. Finally, just as we reached the bottom of the hill, a chiva stopped and the driver insisted that we should get on. We declined (not supposed to take rides from strangers, no money, etc.), but the driver insisted. Finally we decided it might not be a bad thing as we were tired and still had miles to go (with all those wild beasts out there) so we got in the chiva. Soon we arrived in Coco Solito, safe and sound. We thanked the driver, who then left. I have no idea if he even knew who we were. But, you know? He might have. And he might have met up with my dad somewhere and collected the 10 cents due him!

- *Wendy Cotton Corrigan*

A Bigger Nest

In the twenty or so years my family lived in the Canal Zone, we moved to different quarters six times. I blame it on the Chagres water. My parents kept having kids, and we needed ever expanding room to house them all. Starting out in Coco Solito, we were given two adjoining apartments because we needed bed space for six. Even so, we converted the second kitchen into a bedroom for the youngest child. We then moved to newer quarters in Coco Solo. Even with bunk beds for the two oldest, the new quarters proved too small when a new baby arrived. Back to Coco Solito we went, and again we were given adjoining apartments. And again, the spare kitchen was used as a bedroom. After another child was born we moved to Margarita which was really neat for my older brother and me as we got to live by ourselves in one of the adjoining apartments. I guess the housing personnel didn't think that was so neat so we were moved back to Coco Solito. Once again sleeping in the kitchen became necessary. After two more siblings joined our family, we moved again--this time to larger quarters in Coco Solo. Now with six kids, we were back to bunk beds and while "the kitchen boy" didn't have to sleep where a stove and refrigerator used to be, we did have to cordon off space on the porch for his bedroom. Nice in the tropics, right?

- *Gerry DeTore*

Left, Charlotte Bialkowski standing outside her home in Coco Solito in the early 1950s.

Right, Coco Solito buildings project was still under construction when this picture was taken ot 16-month old Beth Bialkowski. *Courtesy of Mary Elizabeth (Beth) Bialkowski Lozano*

Coco Solo

Sponsors

Lester, "Chefa" & Lena Barrows
Ray & Justine Bunnell Family
Gerry DeTore
George, Bobbi, Bruce & Debra Egolf
Andy Nash English
Dennis & Peggy Huff
Philip & Weulcia Wilkins Family

What About Coco Solo?

The Panama Canal Review,
January 3, 1958, pp. 4, 5.

The breeze-swept onetime Navy base, Coco Solo, will become, barring the unforeseen, will provide housing for former New Cristobalites and other Atlantic siders. The opening of a new Company-Government town at Coco Solo which for the several decades spanning the two World Wars was one of the bastions of the United States Navy's forces guarding the Atlantic entrance to the Panama Canal.

In addition to the 175 employees occupying family quarters in the New Cristobal and Fort DeLesseps areas, there are 8 4 Canal families living in one-bedroom apartment buildings who are to be reassigned better housing. Also, there are a number of new employees who have not yet received permanent housing assignments. As a consequence, the evacuation of the New Cristobal area will be only the bigger part of a population shift on the Atlantic side which surpasses any to take place there.

The extent to which permanent community facilities will be established in Coco Solo is still in the planning stage. If the decision is to develop it into one of the principal Canal Zone Civilian towns, buildings and other facilities are available for the reestablishment of services. Among the buildings or facilities available are tennis courts, baseball diamonds, hobby shops, swimming pools and pavilions, and one of the best clubs on the Isthmus. Buildings are also available for refitting or remodeling for use as a commissary, schools, a service center, and a post office. Plans have been completed for the population transfer which is scheduled to begin about the middle of this month and will continue for several weeks. It is expected that 300 or more families will move during this period.

The quarters at Coco Solo do not require extensive renovation and they can be made available for occupancy at a rapid rate. It is planned to move about six families a day and all available personnel and equipment will be assigned to this work during the moving period. Electric ranges will be installed as the quarters are occupied. The Canal administration is considering plans for extensive improvements in the housing facilities in the future. Such items under consideration are electric water heaters, tile floors, and modernization of kitchens. This will be tied in to the overall study of improving the livability of permanent quarters.

The abandonment of New Cristobal as a townsite for American employees will bring a twinge of nostalgia to oldtimers of the Canal organization. The rim of Manzanillo Island has long been a place of residence for them and their predecessors who built and operated the Panama Railroad before them. While the name "New Cristobal" dates back only to about the time the Canal was opened, the history of the residential area span more than a century.

The entire Manzanillo Island became the property of the Panama Railroad under the original concession for its construction. Since the area was never a part of the Canal Zone, New Cristobal was never developed as a town with all facilities for its population and residents there have depended upon the Commissary and Service Center in "Old Cristobal" for their shopping and amusements.

Coco Solo Beginning To Take On Aspects Of Other Zone Towns

The Panama Canal Review, March 7, 1958, p. 6.

The Canal Zone's fastest-growing community—Coco Solo—will be about half grown by the end of the month. By the first of April, Coco Solo residents will be able to restock their food larders, buy a tank of gasoline, get a restaurant-cooked meal, or buy their drug supplies within the confines of their town.

Most of the town's activities will focus around building 100, the big two-story structure which formerly housed the Navy's Post Exchange and sales store on the first floor and a gymnasium on the second floor. Approximately $150,000 will be spent in renovating this building and making the necessary alterations for its use as a community center.

In addition to the restaurant and food-store, a retail dry goods store, shoe store, beauty parlor, barber shop, tailor shop, and shoe repair shop will be located in the building. The dry goods and shoe store will be housed upstairs. The house wares section will be housed in an adjacent building. Part of the plans of the Sales and Service Branch is the establishment of an attractive gift shop similar to the unit recently opened in Balboa.

After installation of these facilities at Coco Solo, both the Cristobal Commissary and Theater are to be closed. The building which houses the theater will be demolished. It is planned to keep the restaurant in the Cristobal Service Center in operation, perhaps on a modified scale, because of the large number of individuals employed in the immediate area and the considerable numbers of bachelors living in the vicinity.

The development of Canal Zone Government facilities at Coco Solo will be at a much slower pace since money for necessary alterations and improvements must come from appropriated funds. The first major Government facility to be made ready will be the elementary school.

It is planned to establish the Atlantic side High School in the new Zone civilian community. The building to be used for the high school will require extensive alterations.

The development of recreational facilities for Coco Solo residents is still largely in the planning stage. This phase of community life in the new civilian town will be developed partly as part of the school program and partly by the initiative of the residents. There are several buildings which can easily be adapted for group activities by civic and fraternal organizations.

Coco Solo Community Center Opened

The Panama Canal Review, April 4, 1958, p. 3.

The formal opening last Tuesday of the big community center at Coco Solo was a festive occasion for Atlantic side residents.

The first floor was altered in time for the opening of a food store, restaurant, and merchandising section.

The tailor, barber, and shoe repair shops will open later.

An old-fashioned community celebration, complete with ribbon-cutting and free balloons, is being arranged for the formal inauguration.

The big building has undergone extensive alteration to make it one of the most attractive units of the Sales and Service Branch. The second floor is scheduled for completion in about 4 months. At this time, transfer of the remaining units of the Cristobal Retail Store will be complete.

Coco Solo High School

The Panama Canal Review, December 5, 1958, p. 2.

Bids to remodel three large former Navy barracks into a first-class modern school plant were advertised in mid-November to be opened December 29. The new Junior-Senior High School will be housed in a large central building with two wings.

The central building will be the heart of the school operations and its central gathering space. The ground floor in this building will contain; the woodshop, metal shop, and the auto-repair shop.

The first floor of this central building will contain a combined school and public library, cafeteria and kitchens, school offices, health clinic, audio-visual and guidance rooms. All of the offices will be air conditioned.

The second floor of the central building will contain an auditorium, seating 614, a music room for the high school choir, four private music practice rooms, a music office and library and space for instrument storage, as well as two dressing rooms. The auditorium will be mechanically ventilated and the other rooms on this floor will be air conditioned.

The three story wing which will be made from present building 1149, on the left as one faces the central building, will have athletic facilities, dressing rooms, and showers on the ground floor. Each of the remaining two floors in this wing will have seven classrooms. A study hall, museum, and lounge will be located on the first floor and the elementary science area, lounges, and a gallery for exhibits on the second floor.

Three classrooms and an office for the ROTC are to be located on the ground floor of the opposite wing. Also on this floor will be an armory and an indoor rifle range.

Five general classrooms, a multi-purpose room, an art classroom, and a home economics and sewing room, will occupy the first floor of this wing. The second floor will contain the physics and chemistry laboratories, a general science area, and space for a biology-science museum.

The new Junior Senior High School building will be located not far from the Coco Solo Elementary School. Work on the latter is now well under way. Target date for the completion of the elementary school is the end of next January.

Gymnasium Planned for Coco Solo

The Panama Canal Review, April 3, 1959, p. 5.

An up-to-date gymnasium which can be used for dances and school parties as well as basketball games and school gymnasium classes will be built this year as part of the athletic facilities at the new Cristobal High School at Coco Solo.

The new gymnasium is to be located adjacent to the new high school at the corner of Conley and Maple Streets in Coco Solo. The building's masonry is being specially designed to give the maximum natural lighting and ventilation. Two gymnasium areas, one for boys and another for girls, will be separated by a folding partition, which can be opened for dances and games. Folding bleachers will be provided for 640 spectators. An exercise area will be located on the mezzanine.

The contract for the gymnasium will also include the rehabilitation of the tennis court and preparation for provision of a football field, a baseball diamond, and a quarter-mile track. There will be outdoor lighting and bleachers for 1,500 persons.

Coco Solo Pool

In The Good Old Summertime

The Panama Canal Review, August, 1966, pp. 12-13.

Hundreds of Canal Zone children are finding enjoyment in the summer recreation program prepared for them by the Canal Zone U.S. Schools Division. The program provides a myriad of activities guaranteed to bring amusement and a good measure of body-building exercise. Gymnasiums and play shelters in all the Zone communities offer archery, tumbling, kickball, basketball, ping-pong, badminton, etc. Summer programs also offer handcraft projects such as ceramics, egg-shell or crushed rock picture making, painting, and making costume jewelry from paper.

Coco Solo Hospital

Yacht Basin Opens at Coco Solo

The Panama Canal Review, May 1, 1959, p. 4.

Small boat enthusiasts of Coco Solo desired to have a yacht basin in their neighborhood. So they proceeded to build one. It was an outgrowth of the Coco Solo Civic Council "Boat and Hobby Shop." The combined efforts of William and George Egger, of the Electrical Division, John Urey, of the Industrial Bureau, and Peter Foster, President of the Coco Solo Civic Council, were instrumental in seeing the building of the Coco Solo Yacht Basin. Much of the cement for the ramp was purchased by members and the work of pouring the cement also was done by the members with the help of equipment loaned by the Dillon Construction Company.

In addition, they cleared the area around the basin and graded it smooth. A breakwater was built to form a calm-water anchorage basin for the use of those members who wanted to leave their boats in the water. However, the building of the breakwater was not an easy venture. A wind storm destroyed half of their breakwater in one night. It had to be rebuilt. Equipment was obtained from the Panama Canal Company to help with the rebuilding. A dump truck was made available by Hauke Construction Company to carry away the debris.

At present the basin has anchorage for deep-draft pleasure craft and a 22-foot deep entrance. Plans for the future include two finger piers large enough to accommodate a 40-foot yacht, the reconstruction of the existing pier, and the installation of tie-up buoys in the basin.

The Coco Solo Yacht Basin was formally inaugurated. Flags flew, ribbons were cut, speeches were made, and small boats sailed in and out of the tidy little basin which was converted from the former Navy Base swimming pool into a safe and convenient harbor for small craft.

Kites Are Fun

The Panama Canal Review, October 1, 1981, pp. 44, 45.

Prizes are awarded along traditional lines—most colorful kite, longest tail, highest flyer, best homemade kite, and so on—but recognition has been given spontaneously in new categories as the situation required.

A "Greatest Tangle of Twine Award" seemed appropriate, for example, when five kites tangled into one line, creating a knotty airborne problem. And "The Flight of the Phoenix Award," named after the bird in Egyptian mythology that became a symbol of resurrection after it self-destructed and then rose alive again, was fittingly awarded to a youngster for his homemade entry. The kite, made with a frame of dry, brittle tree branches, flew for only a few moments before crashing to the ground, whereupon its young designer/builder patiently repaired it and flew it again.

Since 1976 high-spirited Atlantic siders have held a Charlie Brown Kite-flying Contest. The contest's founder, Dr. Mel Boreham, says that kite fliers of all ages turn out for the event, from great grandparents down to, in one instance, a year-and-a-half-old toddler. The event each year coincides with the annual Christmas tree burn in Coco Solo.

During dry season, local residents are given to flights of fancy, when breezes cause the sighing of palm trees and the clacking of bamboo to alert Panama's kiting enthusiasts that there is a good tail wind blowing and it's time to tie one on.

Preparations for the event are not taken lightly. Weeks before the event, participants take notes at the Cristobal High School Library from books on kite building and flying.

Memories

Breakwater

Mom said I could go to the Breakers Club Lagoon to play.

There was a siren's call to the sea where I grew up. Home was a former naval barracks in Coco Solo, surrounded by water on three sides and a swamp on the other. Fishing, wading, and swimming were kid's play. I was the son of a tugboat master and our Panama Canal Zone world existed for international shipping. My buddies and I were weaned in the town swimming pool and cut our teeth in Limon Bay. We developed an exaggerated confidence in the water and arrogance to its dangers.

One summer day in '64, my neighbors, brothers Matt and Willet, were gifted with two small cayucos. These were roughly hewn tree trunks better suited as planters. But to the gang, they were the pride of the Caribbean. We painted them red and yellow.

We stacked the short, stout cayucos on a Radio Flyer wagon to haul them to the Breakers Club lagoon. Rupert followed with an arm full of paddles, a Boy Scout canteen, and a Mary Jane bread bag full of PB & Js. When we reached the shore, we were faced with a choice. To our left, the club's lagoon; calm, green, and feted from poor circulation. It promised safety and a lingering odor. To our right, the clear aqua blue waters of Limon Bay. It surged with life and promised adventure. No contest. We heaved the cayucos onto our shoulders and waded around a cluster of sea urchins to launch them in the bay.

The cayuco wobbled as I climbed into Old Yeller and asked, "Where we going?"

Matt looked across the bay and pointed. "The light tower. There, at the bend of the breakwater."

The cayucos were six feet long. With us in them, they cleared three inches of freeboard. Any wave thicker than a land crab would break over the gunnels. But we had bailing scoops made out of bleach bottles and were ready for anything. Matt and Willet had custom carved paddles. Rupert and I had plywood squares nailed to broom sticks. In our bravado we never considered life jackets. We reasoned that the cayucos were wood and thus unsinkable. Besides, we were all proud owners of "B" for beginner, swimming badges. We gave drowning no more concern than the sharks that fed in the bay.

Willet raised his paddle over his head and shouted, "Yeehaw," every time a rogue wave broke over the bow and sprayed his face. We were well into the belly of the bay where nothing blocked the onshore wind and the waves ran free. More water splashed over the red and yellow gunnels. The wind pushed the bows one way, then the other, making our course a zigzag.

The sea rolled over our dugouts without a care. Rupert was washed over the side and Matt out the stern. Willet was pushed back onto my lap. Both cayucos filled with brine and the sterns settled below the surface. The bows flashed red and yellow in a nod to the Titanic.

Everyone was present and accounted for.

"We got wooden submarines now," I said.

We rolled the hulls over as high as we could to trap air under them. That made them hobby horse buoys and we mounted up. "Yeehaw," whooped Willet.

It was nice to rest my paddle-weary arms but sitting on the hull was no break from the Panama sun. I rolled off, ducked under and surfaced within the hollow of the cayuco.

My buds popped into the air bubble and we played wolf pack.

Invigorated, we set about re-floating our boats. We rocked them fore and aft; slopping most of the water out, then bailed the rest.

There was a concrete block the size of a VW Bug on top of the wall. It had a panoramic view of the bay to the south and Caribbean to the north. The four of us sat on it and washed down soggy peanut butter and jelly sandwiches with tepid canteen water. It was a feast; nothing accents a meal like adventure.

We talked over the whine of onshore wind and the booming percussion of waves on rock. They were loud, big, dangerous, and irresistible. We were all thinking it but Willet called it, "Wet Red Rover!"

The wave roared its intentions as it charged our line. I took a deep breath and leaned into it. Reflex closed my eyes at the last second. It delivered a full body blow. The force of it ground me back into the rock. I concentrated on my grip on Willet. The engulfing mass broke over and around us. It held, hesitated, and then began to withdraw. Clawing water pulled at arms and legs. It pushed from behind and nudged us toward the brink. It seeped under and swept young Willet off his perch. He kicked his feet in the foaming soup to break its grip. Matt and Rupert pulled on his left arm. I planted my feet and pulled on his right. We hoisted Willet back onto the rocks. Our line held. "Yeehaw!" cried Willet.

The gang backed up a couple of boulders and let the high surge pull at our feet before climbing into the cayucos for home. The wind was at our back and the chop carried us toward Coco Solo.

Mom didn't look up from the potato she was peeling. "How was the Breakers?"

"Okay," meaning great.

- Brian Allen

Higher Aspirations

Most of us aspire to get up in the world, and for us growing up in the Canal Zone it was no exception. In the 50s I lived in Coco Solo, then a Naval Air Station. Being such, the water tower in the town was painted with large alternating red and white squares so the tank would be more visible to pilots. One evening, a friend and I decided to climb the tower. My goal was to be personally higher than any point in the town. A secondary goal was to let all Zonians know we had climbed the tower. Upon reaching the platform around the tank we painted our initials five feet high, one with red paint on a white square and the other with silver paint on a red square. One goal completed. I then climbed up the side of the tank and lying on the very top extended my hand above the red flashing light. With that personal goal completed we climbed down. My friend and I were both working as life guards at the Coco Solo swimming pool a few days after our "towering experience" when I saw a Canal Zone police car drive up. I knew right away we were busted. After a ride home and telling my mom what I had done, she told the policeman, "Well, Gerry always did have his head in the sky." My dad had some other words for me when he came home from work.

- Gerry DeTore

Childhood Days in Coco Solo

I lived my childhood days in Coco Solo no more than 200 yards from the Sea Wall. Between a manicured open field and plots of stickers and fire ants and where I caught red snapper on hand line with shrimp. Sometimes as a pay off to some of my good deeds turning sour. Mostly concerning our maid, Dioselina.

I caught some baby alligators from the first bridge along a ditch that headed out to Galita Point. The pay off was grand and getting a good grade in a flunking science class by supplying some animals. On one occasion I put three baby gators (probably crocodiles, too) inside the tub downstairs which were so huge you could fill it with water and do double ganners off the edge and not hit bottom. It never did occur to me that Dioselina would be using that wash basin so soon. When she saw those gators she went absolutely loca. She chased me with words of "matar" running amuck, but this flackito was too fast. Thank Gawd, because I would not be here today. For days it was like trying to evade a jaguar who was constantly stalking its prey. ME. If that wasn't bad enough, my Dad got wind of the incident and I again cheated death by him controlling his very very very mad at me ways. It was like machine gun fire of words. Where did you get these? Do you know these things have mamas? Do you know that the mama is close to her babies? Do you know that they will EAT YOU if they catch you??? The old man was not one to mess with and to speak of it does not give justice as to how he said it. Put it this way: He was not a happy camper.

When we moved to Diablo in the years following this time frame, we would visit her whenever we came back to the Colon side. We would bring her things and food, and she also would come out to visit us in Diablo. I am so glad I never ever let her know about the two boas I had in a cage that got away from me in her room, and I NEVER found them. If I had let her know, I would still be fishing today for red snapper to keep her contended. I cannot end this without saying that this woman was such a lot of fun. Although she chased me many times, I loved her precious soul. And her, me, whenever she put that stick down.

Ahhhhh!!! Mi Panama Canal Zone, my childhood, my friends, my life's most happiest memories kept not in a safe, but branded in my heart forever.

- George Husum

Housing units in the town of Coco Solo.

Coco Solo Teen Club.

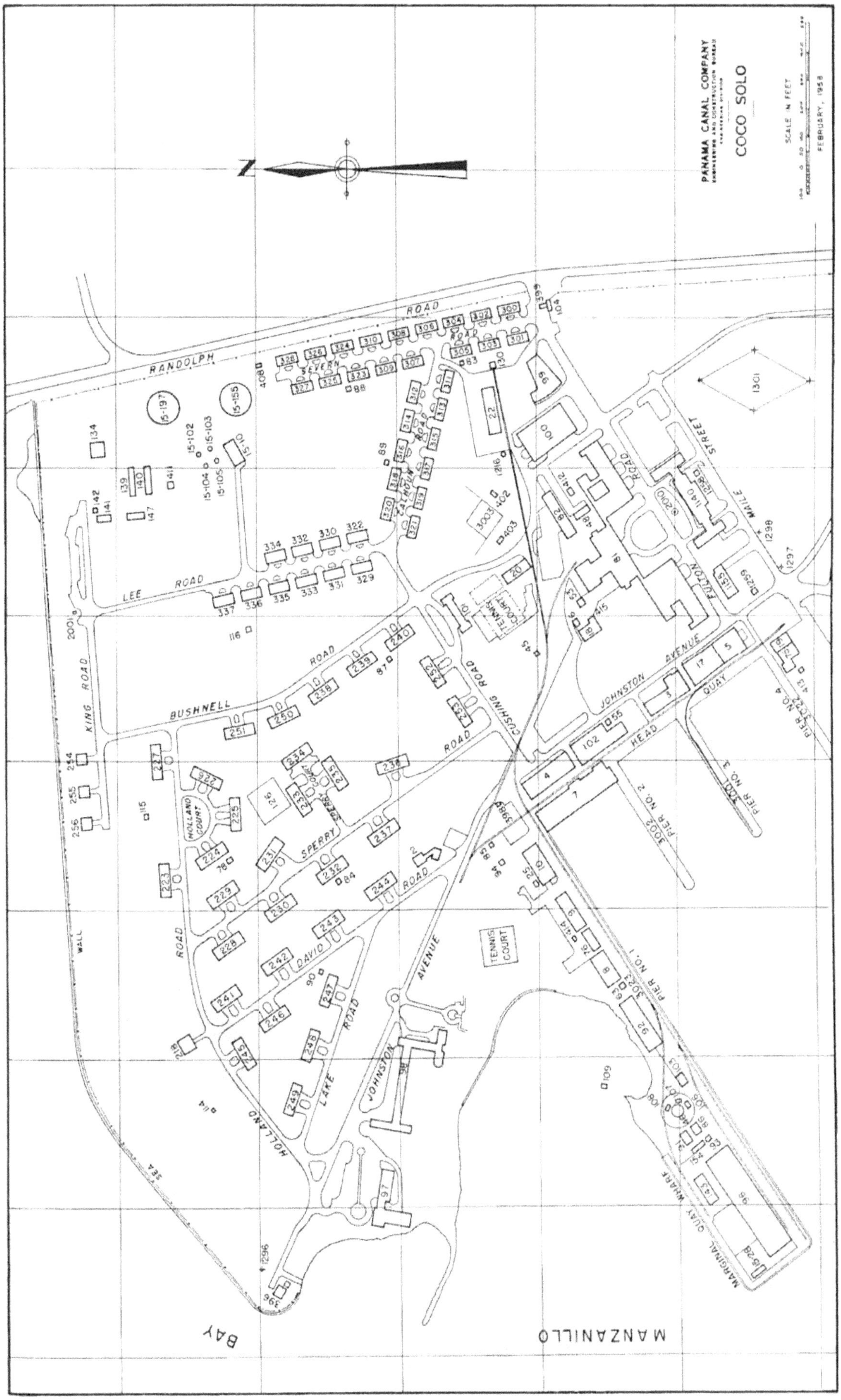

PANAMA CANAL COMPANY
ENGINEERING AND CONSTRUCTION BUREAU
SPECIAL ENGINEERING DIVISION

COCO SOLO

SCALE IN FEET

FEBRUARY, 1958

RANDOLPH ROAD

SEVERN

CALHOUN

LEE ROAD

KING ROAD

BUSHNELL ROAD

CUSHNING ROAD

HOLLAND COURT

SPERRY ROAD

DAVID ROAD

LAKE ROAD

HOLLAND ROAD

JOHNSTON AVENUE

JOHNSTON

WALL

SEA WALL

TENNIS COURT

MAILE STREET

FULTON ROAD

HEAD

QUAY

PIER NO. 1

PIER NO. 2

PIER NO. 3

MARGINAL QUAY MOLE

BAY

MANZANILLO

First Quarters Construction Period - 1906-1920s

Screens were added to windows on this renovated French era home in Corozal.

Most of the housing stock left over from the French construction era, like the one on the right, were claimed by the jungle, some of which could be restored for American workers.
From the Hallen Collection.

Left, A two-family Type 4 quarters in Culebra in 1910. *From the Hallen Collection.*

Right is one the 12-family units, Type 201, at Williamson Place in Balboa.

This photo taken in 1916, Ancon shows one of the first concrete block quarters in the Canal Zone. *From the Hallen Collection.*

Cristobal

Sponsors

The Joseph & Carol Coffin Family
Brian Cox
Donna Geyer Cox
Kathleen Cox
Kevin Cox
Larry Cox
Lynda Geyer
Dennis Huff
Patti Maedl Krough
Michael & Elaine Stephenson Family
Family of James & Stacia Walsh

New Cristobal

Sponsors

CHS Class of 1949
W. R. Dunning Family, Bill, Pat, Danny, Sandi & Vicki
Virginia Kleefkens Rankin
Joseph H. White Jr., & Family

Old Cristobal

Sponsors

Lester, "Chefa," Charles & Lena Barrows
W. R. Dunning Family, Bill, Pat, Danny, Sandi & Vicki
Joseph H. White Jr., & Family

Your Town - Cristobal

The Panama Canal Review, October 7, 1955, pp. 8-9.

CRISTOBAL looked like this in the "good old days." The caption on this old photo says, merely, "Cristobal before paving, 1907." The trees bordering the unpaved streets are coconut palms.

Cristobal is the Canal Zone's front door and its kitchen door, too.

Most of the Canal Zone family-the people who live and work here-entered their new home through that front door and if they didn't, their fathers or grandfathers did.

Guests have come through that front door, too, on a good many occasions. Presidents, Cabinet Members, Congressmen, and even Queens and Dukes have seen the Isthmus for the first time from a ship entering Cristobal harbor.

The kitchen door is where the groceries are delivered. A good part of the food every Commissary-supplied Zonian consumes starts its trip to his table from Cristobal. Before the Canal was opened in 1914, most of the material to build it and the supplies to keep it running were unloaded from ships docking at Cristobal. Even today, three times as many ships dock at Cristobal as at Balboa.

Manzanillo Island

The geographic layout of the Atlantic side is confusing, even to those who have lived here for a long time. Cristobal, Colon, and New Cristobal are all on Manzanillo Island, connected to the Isthmus of Panama by the fairly narrow strip on which the Motor Transportation Division now stands. There is not now a swimming beach along the whole shoreline except at Fishermen's Village in Colon.

The island was a virgin swamp of about 650 acres when the Panama Railroad selected it for its headquarters in 1850. On Manzanillo Island, the Railroad Company established its offices, its docks, its railroad terminal, and quarters for its officers and employees.

When the French forces first arrived in Panama in the early 1880's, they found the Railroad Company, as first comer, in the choice site along the northern tip of the island. Colon was only a few streets wide and long and the rest of Manzanillo Island was still a swamp.

Throughout the history of the Isthmus, land has been made when it was not naturally at hand, and this the French did.

Next to Colon, on a coral reef, they dumped spoil from their canal and on this artificial plateau they built their warehouses, shops, roundhouses, office buildings and quarters. They named this section—from about the present railroad tracks to the old Fort Sherman ferry slip—Christophe Colomb, or Christopher Columbus. It was an easy step from the French Christophe to the Spanish Cristobal.

By the time the American Canal builders reached the Isthmus, Manzanillo Island was divided, like Gaul, in three parts: The Panama Railroad area from the old freight house to the section where the empty windows of the now unused Colon Hospital look out to the Caribbean; Colon, inland from the railroad tracks toward the marshy center of the Island; and Cristobal, the French village.

"Old" Cristobal

The Canal Commission of 1904 set up provisional headquarters in Cristobal "in the buildings erected for the residence of Mr. DeLesseps." Mr. DeLesseps was Charles, son of Count Ferdinand DeLesseps. These buildings were on Cristobal

CIVIC COUNCIL President Virgil C. Reed heads the group representing Cristobal and their suburban neighbors from the town of Margarita.

Point, facing the sea; oldtimers place them at the very end of Roosevelt Avenue a few hundred feet from the present Sea Scout shack in Old Cristobal. The statue of Columbus and the Indian maid, later in the courtyard of the Hotel Washington, and now on Broadway between 2d and 3d streets, stood on a knoll between the house and the sea.

Cristobal at first was a municipality, one of five in the new Canal Zone. It was bounded by the sea on the north and stretched along the Canal line to include Mount Hope and Mindi. One of its early mayors, from October 12, 1905, until the municipalities were abolished in 1907 was M. C. Rerdell, later District Judge at Cristobal.

Because of Cristobal's importance as the port of entry for construction equipment, high priority as given to construction there; Master Builder W. M. Belding was constantly under pressure for faster work. At one time Chief Engineer John F. Stevens warned him that "we are already cramped there (Cristobal) for room and must immediately stop bringing any more white men until additional quarters are provided."

1906 -1907

By April, 1906, Cristobal had a population of 2,010—489 of them Americans. The Cristobal Club had been organized the big clubhouse at Columbus and Broad Streets came later—and given temporary quarters. There was a postoffice which handled mail for the West Indies that formerly had been routed through New York. An eight-room school house was being built and brick-and-cement jail, amply commodious to meet all existing requirements," was accepting its first involuntary guests. Its commodiousness was happy foresight; for years, the number of arrests in Cristobal was higher than in any other Canal town.

The Cristobal Fire Department was one of two paid companies in the Canal Zone—the other was in Ancon—and the offices of the Fire Chief were at Cristobal. The department's hose wagon, chemical engine, hook and ladder truck, and steam engine were local wonders.

Cristobal was never headquarters for a construction division but it was headquarters for the Division of Materials and Supplies, something like today's Supply Bureau. Its bakery, cold storage plant, and warehouses lined one side of Cristobal Point. The present Sea Scout shack and the big masonry building adjacent to the Maintenance Division yard were commissary structures. The concrete bases of others are barely visible in the tall grass along the bayside.

Cabbage Patch

Most of the officials, like W. G. Tubby, Chief of the Division of Materials and Supplies, and Lorin C. Collins, Associate Justice of the Supreme Court, lived in houses along the waterfront west of the DeLesseps house. In the less choice section further inland were four-family houses and bachelor quarters. One neighborhood was known as the "Cabbage Patch." An early resident complained to the Canal Record that wandering goats ate up his rosebushes.

A commissary "for the exclusive use of gold employees" was opened on the site of the present commissary about December 1, 1907. According to the Canal Record it dealt in "mens' furnishings and articles for ladies" and had a "waiting room for ladies from the Line in which they will be made comfortable until the departure of the trains."

Social life in Cristobal in 1907 was typical of that of the larger Zone towns: by that time Cristobal's population was over 4,000, a quarter of them Americans. Lodges were important and numerous. The Improved Order of Red Men, then the largest fraternal order on the Isthmus, had a Tribe in Cristobal, and the Knights of Pythias organized their first local lodge there. The Elks and Masons were both active. For the women, there was the Cristobal Woman's Club; organized September 27, 1907, it is still functioning and has the longest continuous life of any woman's group in the Canal Zone.

The Clubhouse had a first-rate basketball team and Nelson R. Johnson, Clerk of the Circuit Court and once a professional acrobat, offered lessons in tumbling. The July 4 celebration that year featured such events as a ragamuffin parade and a fat man's race. The winner of a greased pole climb was rewarded with a $5 gold piece—it was glued to the top of the pole.

That was Cristobal, once a French village.

"New" Cristobal

Meanwhile, in the old Panama Railroad area at the tip of Manzanillo Island things were humming. Before the American canal forces arrived, there had been two hospitals on the beach—the Railroad's 30-bed hospital for its employees and the larger French hospital which adjoined it. The railroad hospital dated back to the early railroad days, the French hospital to 1883.

These two were combined as soon as the Americans began work on the canal. New buildings were put up and by November 1906, the hospital consisted of 40 frame buildings. Five of these stood on brick and concrete pillars on a coral reef which was covered by the sea except at low tide, and the rest were in a 35-acre plot with a sea frontage of about a quarter of a mile.

The first school on Colon Beach— it had 38 pupils—was opened early in February 1908. Up to that time the only school available to the younger children of American employees of the Railroad had

CAPTAINS W. Casswell of the Fire Division and Eugene S. Shipley of the Police Division are responsible for policing and fire protection of Cristobal.

been the Cristobal school, a good distance away in that pre-motorized age.

In this section were four rows of quarters near Christ Church, which was built In this section were four rows of quarters near Christ Church, which was built by the Panama Railroad in 1865; other quarters, which had been made from a remodeled storehouse; and Garfield, McKinley, and Lincoln Houses which at the time sheltered both families and unmarried employees. Lincoln House is gone but the boarded-up shells of Garfield and McKinley Houses are still standing. The original Washington House, later the Washington Hotel, was the fourth of this residential group. It was both a lodging house and "eating house" for railroad employees long before 1904. It was replaced about 1913 by the present Hotel Washington.

Not far away, about where the present elementary school stands, was the four story, brick Mechanics' Building, a swank apartment house of its day. Mrs. Ida May Cotton, a real old timer, remembers living there when she was about 12 years old. The Mechanics' Building was torn down in the 1930s.

Two major changes came to Cristobal about the time the Canal was opened to traffic. Work began in earnest on the great piers which today protrude like out-spread fingers into Limon Bay. This meant, among

CRISTOBAL'S school principals are Mrs. Helen L. Rushing for the elementary school and Paul L. Beck for the senior and junior high schools by the sea's edge.

other things, that shipping companies moved from the freight-house section into Cristobal proper.

The other major change was the construction of Fort DeLesseps, a Coast Artillery post, on the shoreline of Manzanillo Island between the freight house and the Hotel Washington. This meant the loss of what had been a residential area and led to a revision of the entire Atlantic side town planning.

Cristobal itself was set aside primarily for a commercial section-the center of civil administration for the Atlantic side and headquarters for shipping agencies, banks, freight handlers and the multiple activities of a busy port. The location of a residential section was not so easy to decide. The Canal administration finally ruled against a townsite at Mount Hope and decided to expand the first Manzanillo Island settlement—the Panama Railroad section along Colon Beach.

This meant a great fill, to build up the swampy heart of the island, and eventually the new fill stretched from Melendez to Roosevelt A venues and from 3d to 9th Streets. Along the waterfront the old radio station and quarantine station disappeared, new hospital buildings were constructed, an elementary school (1918) and a new high school (1933) appeared. Quarters were built gradually from 1917 through 1938. Today the population trend is away from Manzanillo Island and toward Margarita and many of the New Cristobal houses are vacant. Under the terms of the new Treaty, the United States agrees to seek legislation for transfer of these lands to Panama.

Cristobal in Wartime

No story about Cristobal could be complete without mention of the part it has played in two world wars. During World War I, censorship headquarters for the Canal Zone and Panama were established above the Cristobal postoffice. Servicemen from the new Army posts at Fort Sherman, Fort DeLesseps, Fort Randolph and the Coco Solo Naval Base jammed the streets and filled the commissary and clubhouse. Atlantic siders turned out each week for Liberty Bond rallies and there was a War Savings Stamp drive, headed by Emmett Zemer, then working at the Hotel Washington.

From the hotel's porch, guests watched the Navy stretch a submarine net across

HOUSING MANAGER, Wendell G. Cotton, and Postmaster Otto L. Savold are in charge of shelter and mail for residents of the Cristobal area.

COMMISSARY MANAGER, O.W. Ryan and Service Center Manager Joseph Pustis provide the meat and milk and movies for Cristobal.

the breakwater entrance each evening at 6 o'clock. Ships caught outside had to wait until it was removed the next morning. Convoys of ships, their hulls painted in stripes or mottled like World War II's jungle uniforms, formed in the harbor.

Cristobal women were angels of mercy for troopships bound to and from New Zealand and Australia and Europe. They took the "walking wounded" on sightseeing trips, ending up for luncheon at the old Gilbert House or dinner at the Hotel Washington.

War Came Close

Twenty-three years later, during World War II, war came much closer to Cristobal. There were the crowds of servicemen, the food shortages, the convoys, this time of solidly gray ships. But this time there were blackouts. Hotel guests stumbled along darkened corridors to the dim light of a kerosene lantern turned low, and every room displayed directions for the quickest route to the air raid shelter under the building. Every few days Fort DeLesseps' anti-aircraft batteries, on a revetment above the swimming pool, shook the Washington's west wing.

One dreadful month, in July 1943, Cristobal's streets were full of survivors from ships torpedoed in the Caribbean.

Some had gone down only a few hours out of Limon Bay. And one morning, August 19, 1942, a Navy seaplane, taking off, crashed into a Panama Canal tug, killing practically everyone aboard. Men and women watched the flames helplessly from along the beach. J. C. Randall, now Chief of the Housing Division, was one of the horrified witnesses.

Today, Cristobal is far more a center of commerce than a residential community. Its population in last November's census was 562—119 of them Americans. New Cristobal, with a total population of 1,130, was far smaller than it had been.

Nine steamship companies have buildings in Cristobal, all clustered around Steamship Row. Practically all of them combine office space on the ground floor and living space above. Handily nearby is the All America Cable Company's Atlantic side office.

The Veterans of Foreign Wars and the American Legion have meeting places in Cristobal, but the latter is talking of giving up what was once an old church to move to Margarita.

Cristobal Today

The people of Cristobal—the old French village—may shop, select books at

their library, consult the housing manager, attend movies, mail letters, buy tickets for police or firemen's balls, be brought before the courts of justice, all without leaving their home base. But for church going, with very few exceptions, for social life, or for sports they must go elsewhere. Their sole club, and its membership is composed of small-boat fanciers, is the Panama Canal Yacht Club where the steaks are famous.

Modern Cristobal has a distinction enjoyed by no other Canal Zone community. It has a coconut collector. The law of gravity never having been repealed and falling coconuts being potentially lethal, a concessionaire makes frequent rounds harvesting the ripening nuts.

Cristobalites share with other Atlantic siders a philosophy of their own: They don't see why anyone ever wants to live anywhere else. They resent, quite vocally, being considered "second-class citizens" as they feel they are sometimes treated. They yield to no one in the defense of their community, their pride in the sparkling blue bay and its palm fringed shore, their great piers and the trade winds of the dry season. They are Cristobalites first, Zonians second.

60th Birthday For Cristobal Woman's Club

The Panama Canal Review, November 1967, pp. 9, 21.

The U. S. Government, concerned with the problems of the building of the Panama Canal, was not too busy to consider the morale of the wives of the employees. Miss Helen Varick Boswell was sent from Washington's National Civic foundation to look into conditions. She believed that a woman's club would be the answer.

The Cristobal Woman's Club was the first of a group of seven Isthmian

clubs formed 2 weeks later, September 27, 1907. Under the name of the Canal Zone Federation of Women's Club, these clubs were landmarks of the early construction days of the Panama Canal. Of the original clubs formed under the Canal Zone Federation, the Cristobal Woman's Club is the only one in continuous existence from the day it was founded.

The old YMCA building in Cristobal was the Club's first meeting place. From 1917 to 1952 the Gilbert House was Club headquarters. The American Red Cross gave the Club permission to use their building from 1952 to August 1959, when the Cristobal Woman's Club building was completed in Margarita.

Tuesday morning workshop with members busy in Bazaar preparations.
The Panama Canal Review, November 1967, p. 9.

The Club's philanthropy program has been active since the Club was formed 60 years ago. Each Thursday of every week in the year, food and clothing are distributed to about 80 aged indigents of Colon, most of them over 75 years of age.

The 60th anniversary of the Cristobal Woman's Club was observed with an address by Gov. W. P. Leber on the Canal's future. Mrs. Harry Butz, president, presided and, symbolic of the Club's history, used a cocobolo wood gavel that dates back to French construction days and which was salvaged from the attic of the old Panama Canal Administration Building in Panama City many years ago.

The Cristobal Woman's Club has been in its own building in Margarita since August 1959. *The Panama Canal Review*, November 1967, p. 9.

60th anniversary meeting; Governor W. P. Leber and officers of Cristobal Woman's Club.
The Panama Canal Review, November 1967, p. 21.

Cristobal Woman's Club enters 83rd year

The Panama Canal Spillway, Friday, October 13, 1989, p. 4.

In a broad sense, the seven Isthmian clubs formed under the Canal Zone Federation of Women's clubs contributed to the building of the Canal. The theory was that as long as the women were happy in cultural, social and philanthropic pursuits, happy too, were the men digging away at the Canal.

As the Canal neared completion, the Federation of Women's Clubs dissolved. The Cristobal Woman's Club, however, survived this period, and even though some others later reorganized, it is the only one still alive today.

Holding the club intact today, as throughout its history, is its Philanthropy Department, founded in 1912. Through its philanthropic arm, the

club rolled bandages in World War I, assisted the Red Cross and entertained soldiers who were returning home. In 1921, it opened a Free Clinic, and with the Depression, added a soup kitchen. During World War II, club members made surgical dressings and entertained U.S. servicemen stationed in Panama. Currently, the club serves 50 Colon residents by providing clothing and food staples. During holidays, the club helps with extra food and gifts.

Members of the Cristobal Woman's Club were among this group which met at Empire almost 50 years ago.
The Panama Canal Review, September 6, 1957.

65

HOTEL WASHINGTON

The Washington House/Hotel Washington

The Panama Canal Review, January 1, 1954, p. 4.

The original Washington House was built about 1870 as a residence for employees of the Panama Railroad Company. It was a two-story frame building located on the present hotel site. In 1905, when construction of the Canal swelled American forces in the Canal Zone, a third floor was added to the hotel and in 1908 it was taken over by the Isthmian Canal Commission.

President William H. Taft, a frequent visitor during construction days, was convinced of the need for a good Atlantic side hotel. In 1910, he authorized construction of a new Hotel Washington. The old hotel was moved, the seawall reinforced, and work begun on the new building. On March 13, 1913, the Washington—as we know it today—housed its first guest, a well known American named Vincent Astor. Ten days later it was opened formally to the public.

Operations of the 40 year old Hotel Washington in Colon were transferred today to the firm of Inversiones Motta, S.A. On December 22, Arturo Motta, representing the five-brother company, signed a lease for the hotel and a contract for its management. Governor

An aerial view of the Washington Hotel with its swimming pool to the right of the photo.

Seybold, in his capacity as president of the Panama Canal Company, signed for the company. Thus begins another chapter in the history of the stately-looking, palm-surrounded big buildings facing the Caribbean. It has been a social center for Atlantic siders for many years.

In a statement issued at the time of the lease signing, Mr. Motta expressed the hope that the Atlantic side community would use the hotel facilities and make it their own center to an even greater extent than they have in the past.

New construction of 4-family units, above, and cottages, right, in New Cristobal in 1919. *From the Red Hallen Collection.*

Family and friends gather at the Hanna home in Cristobal in 1948, after Bob and Betty had returned from a vacation in the States, arriving on the *S.S. Ancon.* Upon returning, friends would get together to discuss what's new Stateside . . . latest cars . . . latest fashions . . . new dances and so on.
Courtesy of Cheryl Russell.

In September 1904, the Isthmian Canal Commission authorized the creation of a school system. The first course of study for the Canal Zone schools was developed under the leadership of Superintendent of Schools David A. O'Conner (1905-1908). He also instituted a system of accountability in the areas of attendance and school supplies. A basic principle of the school system was that a student completing work for any grade in the Canal Zone would be prepared to enter the ensuing grade in the United States. The schools for the children of United States citizens functioned as much as possible like schools in the United States, using teachers recruited from the United States as well as books, furniture, and supplies from the United States.

Cristobal Public Free School, 1910. *From the Hallen Collection.*

Above, Cristobal Junior - Senior High School, 1965. Right, Cristobal High School, circa 1938.

 *Memories*

CHS Memories

Remember the courtyards in the CHS building before the school was moved to Coco Solo? Along with that huge old turtle that roamed the yard "forever," you could occasionally find sloths, iguana, boas, and just about any animal caught by a student while exploring the jungles around our homes. Myself and another senior even tried to place an underclassman in the courtyard once, but we were caught in the act by Ms. Dorothy Smith, the English teacher as we tried to drop him through the classroom window next to the yard. Needless to say, she was not pleased.

The Cristobal Elementary and Junior/Senior High School were each located right next to Manzanillo Bay. Sometimes it was difficlult to concentrate on our studies while watching the sea planes from Coco Solo and France Field land and take off on the beautiful waters of the bay. The water was so clean, some of us kids would often get in 15 or 20 minutes of swim time during our school lunch break, frowned upon by school officials but we did it anyway. We either swam in the bay or walked the coastline picking up interesting shells or popping the air bags of Portuguese Man of War jellyfish that washed up on shore.
- *Gerry DeTore, Class of 1959.*

Monkey

When I was a little girl, maybe five, we lived in Cristobal, and our neighbor—who lived, oh, two or three doors down-- had a monkey cage in his back yard. As I recall, it was really big, like ten feet high and five feet square. Of course, it had a monkey in it. Well, the kids used to come and tease it and poke at it, and the monkey grew mean.

One day I went outside to empty the garbage. I looked down the street's hedge, and I saw that monkey glaring at me. I RAN up the steps with that monkey on my tail. He scampered right into the house and cornered me, cowering and crying, in the kitchen sink, and he began to bite my legs. My mother, meanwhile, had run to my baby sister's room to shut the door. By the time she came back I had been bitten half a dozen times. I still wear the scars from that attack.

We went to the emergency room, and there the nurse began to swab my wounds. I was sobbing my little heart out when she said, "Why are you crying? It's only soap and water." I toned it down to a whimper.

The morals? Never trust a monkey, and to this day, I don't.

And, don't cry over soap and water. To this day, I don't.
- *Edna Hart Crandall*

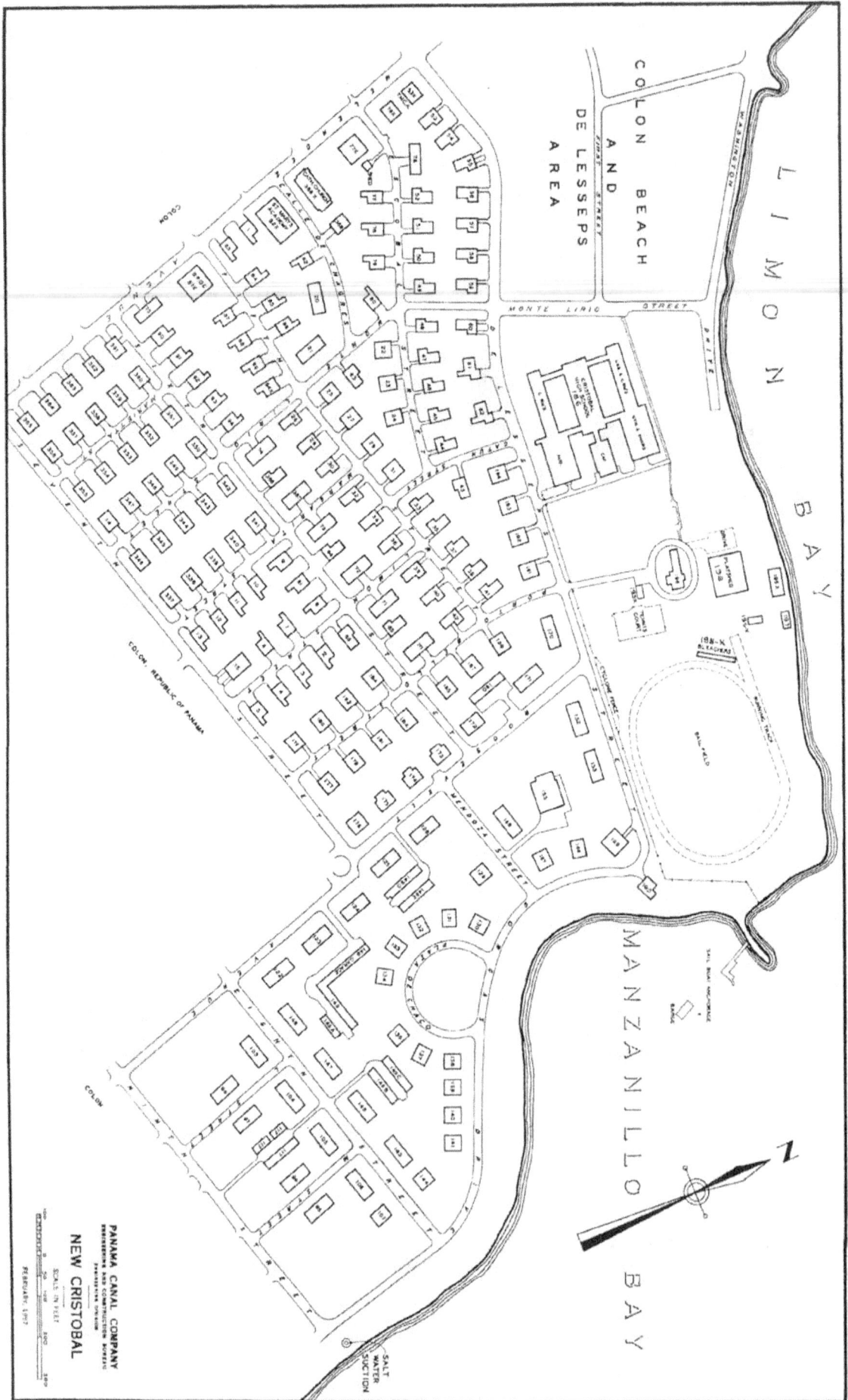

COLON BEACH
AND
DE LESSEPS
AREA

LIMON BAY

MANZANILLO BAY

PANAMA CANAL COMPANY
ENGINEERING AND CONSTRUCTION BUREAU
ENGINEERING DIVISION

NEW CRISTOBAL

SCALE IN FEET

FEBRUARY 1957

Diablo

Sponsors

Baglien Kids: Lynn, Beth, Julie, & Joel
Katherine Egolf
Andy Nash English
Gregory Stephen Gramlich
Dennis & Peggy (Hale) Huff
Charles W. "Chuck" Hummer
Pat & Jack Hunt, Lisa, Terri, Laurie, & John
Joan McCullough Ohman
Roy, Peggy, Jim & David Reece
Bob & Cheryl Russell
Michael & Elaine Stephenson Family
The Wards—Marvin, Jackie, Debbie, & Lissa

Diablo Heights

Sponsors

Connie Glassburn Dawson
Paul Glassburn
Val & Bert Schroeter Family (Suzie & Tina)

Your Town - Diablo Heights

The Panama Canal Review, March 5, 1954, pp. 8-9.

DIABLO HEIGHTS was a town of 12-family houses when this picture was taken about 10 years ago. Today not a single one remains.

Although it is one of the newest of the Canal Zone's permanent towns, Diablo Heights--as far as its history is concerned--may antedate almost any other existing town. Like a phoenix, it has risen from a succession of ashes, figuratively speaking.

During the middle part of the sixteenth century, according to Isthmian histories, the narrow Isthmus of Panama was terrorized by bands of *Cimarrones*, runaway Negro slaves, who preyed upon the treasure trains on the *Camino Real*.

They became such a threat to life and property that the Spanish viceroy sent expeditions to clean them out. They evaded such punitive parties and in 1552 were granted recognition by the Governor of the Province. At that time they had three main villages, one of which was called Diablo. It was located near the site of present day Diablo Heights.

The saga of modern Diablo began early in the twentieth century. Marine detachments, which came to the Isthmus during Panama's revolutionary period, were stationed at Diablo, as well as at Mount Hope and Empire. Later, in 1905, the Marine headquarters were established at Camp Elliott, on the west side of the Canal not far from Culebra and Empire.

First A Labor Camp

Late in 1906 members of the Isthmian Canal Commission began to consider locations for labor camps. These were to be located as close as possible to the places where work was going on, or where it was expected to go on. One of these was to be at the "north end of the proposed Sosa-Corozal dam, on high ground south of Corozal as near the end of the proposed dam as possible." It was to house at least 200 laborers and was to have five barracks, a kitchen, a messhall, and a "gallego" (gallegos were European laborers) messhall and kitchen.

In May 1907, Chief Engineer George W. Goethals officially named the labor camp "Diablo Camp." There is local tradition that during the time it was used as a labor camp, Camp Diablo was the scene of a bloody "wine riot." European laborers had been issued wine with their meals on holidays. When the regular issue failed to arrive they rebelled and bombarded anyone who ventured their

way with empty bottles, sticks or stones, or anything they could find to throw. As the story told by old timers went, a padre was called on to settle the fracas as he was the only man who could get near enough the enraged laborers to negotiate with them.

When the Sosa-Corozal dam project was abandoned, the six laborers' barracks at Diablo were converted to quarters for 12 American families. Diablo was connected with Corozal by road and was considered part of the town of Corozal. These old barracks-quarters were occupied until the end of the construction period.

In 1914-15 the Panama Railroad, which had run from Corozal to Panama along a route which, roughly, crossed the upper end of what is now Albrook Air Force base, was relocated so that it would run into Balboa. The relocation meant that part of Diablo Hill had to be cut away. The last two of the old buildings on the hill were vacated January 5, 1915.

Several Plans Considered

During the 1930's there were several plans made and unmade for the Diablo site. One called for a local-rate town there; another proposed Diablo as a new location for the oil-tank farm at La Boca. A third project, quickly rejected by the Governor, had to do with establishing a private, noncommercial flying field on the Diablo fill where spoil from the Canal channel was occasionally dumped.

It was not until July 1939, that a definite decision was made. Diablo Heights--although it was still being referred to as Camp Diablo--was selected as headquarters for the Special Engineering Division, with two office buildings for the planning staff and quarters for about 280 employees. The office buildings, two 32-room bachelor quarters, twelve 12-family houses, 65 cottages, ten 10-family buildings and garage buildings were all to be built with funds allocated to the SIP, or Special Improvement Projects which had to do with the installation of protective devices on the Locks.

The fast-growing force, however, called for roofs over more people's heads and additional barracks and more 12-family houses were added.

In the meantime Diablo Heights still had no official name. Early in 1940 there was a spritely exchange of official correspondence on the subject. Crede H. Calhoun, then Chief of the Division of Civil Affairs, added his bit: "We have a Paradise at Paraiso; I see no reason why we should not have a Devil at Diablo." Governor Clarence S. Ridley suggested that the name might be Diablo Heights.

Town is Officially Named

Postal people pointed out that there might be confusion with so many "Heights," Quarry Heights and Balboa Heights being names of considerably longer standing. C. A. McIlvaine produced the idea that Diablo Hill might be hispaniolized into "Cerro Diablo," but the Governor held firm to his original stand. On February 5, 1940, the reborn town was designated officially as Diablo Heights.

A commissary was opened in May of that year and the post office the first of the following month. The Diablo Heights school began sessions in September 1940, with 210 pupils. Work on the clubhouse was begun in October and the building was officially inaugurated the following June.

In the meantime Diablo bachelors--and others--had their meals at the "restaurant," a long, one-story building which stood near the present intersection of Walker Avenue and Endicott Street. It soon became known as the messhall, a designation which has been carried over--even by teen-agers who never patronized the original messhall--to the present clubhouse. Open 24 hours a day, it was crowded at almost any hour and became a favorite place for after-party coffee and sandwiches.

Diablo Heights in those days was a lusty place. A good many of its residents were construction people; they brought construction camp customs and traditions with them. Some of them rebelled against what they considered the "stuffiness" of the Canal Zone; family brawls and neighborhood drinking parties were a constant headache to the police.

But drinking and brawling was not all they did. They worked and they worked hard. After hours they turned to making their own fun. There was a Diablo Heights Recreational Association, a local newssheet, and a dance club named the

GEORGE L. CAIN, Commissary Manager. (This really is. His brother, Sgt. Edwin B. Cain of the Pedro Miguel Police, was identified wrongly in last month's issue.)

MISS RUTH E. CREASEY, School Principal

WILLIAM T. HALVOSA, Postmaster

aluminum houses. Other permanent houses of new types--the idea was that a pair of permanent houses was to have between them a temporary house during the crowded construction period, the temporary quarters to be removed when they were no longer needed--went up in Diablo. All of these are still standing in Diablo Heights and always baffle newcomers.

Diablo Heights, in this incarnation, however, was never the uproarious place it had been during Third Locks and SIP days, and it lapsed into a near coma when the sea-level canal studies were finished in 1948.

Excess Housing Rented

During the subsequent comatose period there was so much excess housing in Diablo Heights that the Canal was able to offer 12-family accommodations to servicemen and their families. In 1950-51 there were 207 service families--most of them Army--living at Diablo Heights. They were all required to vacate the quarters when demolition of the 12-family houses began.

When the housing replacement program was started in 1950 Diablo Heights, with its undeveloped sections, was a logical choice for some of the new construction. New streets were laid out, hilltops bulldozed off, and quarters began to go up. The first on-the-ground-masonry quarters were built there. Today more new quarters are under construction where the 12-families once stood, and the town's streets are again being rearranged.

Today the population of Diablo Heights is again on the upswing. Units of

MELVIN E. WALKER, Clubhouse Manager

the Comptroller's Office have moved into the three-story office building; the Payroll Branch has been at Diablo Heights for a number of years. From a low of 805 in 1951, the figure has increased until last March it was 986-almost equally divided between men, women, and children. (The dog population is uncounted, but from all indications it is high.)

A Wholly Residential Area

There is no church in Diablo Heights. The Camera Club, which occupies an attractive, well-kept building on Hains Street, is the only organization housed there. The Diablo Heights Dispensary was closed some time ago, and the fire station last year. Police and fire protection is furnished from the Balboa Central Stations. Diablo's Civic Council is one-third of the tri-partite Pacific Civic Council.

Except for its office buildings Diablo Heights is a residential suburb. Its residents are clerks and pilots, teachers and engineers, electricians and boilermakers, carpenters and office workers. But almost without exception all of the "blue collar people" work elsewhere; the small Dredging Division unit near the Canal bank has the only craftsmen regularly employed in town.

Diablo Heights is cool, drowsy, and, except for the motorcycle enthusiasts who make the nights hideous with their noise, quiet. The clubhouse is open all night and early risers like customs inspectors or railroad men due for an early call get breakfast there. But most of its house lights are out well before 11 o'clock. The days of wine riots of the early 1900's and the lusty brawls of the 1940's are behind it--for a while at least.

"Forty Club." Forty is short for 1940, the year when it was organized.

Sidewalk Meeting

Early that year, the Balboa Women's Club had given several get-acquainted dances for the Diablo Heights newcomers, who soon conceived the idea of forming a club of their own.

"So," according to Mrs. Ella Wertz, now of Ancon, who was a charter member of the Forty Club, "we called a meeting for the Balboa movie theater. When we got there the building was locked and we held our meeting on the sidewalk. We gave our first dance a short time later and after that we had dances several times a month at the Tivoli or the Golf Club. Our last dance was the night before Pearl Harbor. What with blackouts and all, no one was much in the mood for parties after that."

As the war continued, shortages of ships, men, and materials required suspension of the Third Locks Construction. As a result the office building on Diablo hilltop began to empty and the population to dwindle. In 1941 its population was 2,003, but 10 years later it had dropped to 805.

There was a spurt of activity for about two years when Col. James Stratton and his force began The Isthmian Canal Studies. Again the commissary and clubhouse were crowded with newcomers and the office buildings were abustle with activity.

Out of this period came some of the strangest buildings ever seen in the Canal Zone. Engineers studying housing for the force which would be needed if third locks or a sea-level canal were built brought to the Isthmus several prefabricated

MOVING DAYS STARTING FOR EMPLOYEES ASSIGNED TO NEW QUARTERS IN ANCON AND DIABLO HEIGHTS

The Panama Canal Review, November 7, 1952, p. 1.

The first five apartments completed were to be occupied this week. The first to be occupied were a masonry duplex, located close to the site of the former Diablo mess hall, and three cottages (a revised version of the Breezeway type, a type 327 with two bedrooms and a type 331 with three bedrooms) nearby. A second group of the Diablo houses is to be turned over to the Housing Division soon and these will probably be occupied by November 15th. All the new Diablo houses will probably be occupied by mid-December.

Type 331 with a living area of 1,211 sq. ft.

Housing Program To Provide Homes For 197 Families

The Panama Canal Review, September 4, 1953, p. 13.

Twelve-family apartments, built at Diablo Heights during the pre-war and wartime boomdays, are fast disappearing from the landscape. Most of the buildings had been occupied originally by Third Locks forces.

All of them have now been sold and are being demolished, as is this quarters building above. New masonry houses will be erected in place of the old buildings.

Thirty-four of the 12-family apartments were constructed in Diablo Heights. One of these was remodeled several years ago into a three-family apartment building.

Type E Diablo Pre-Fab is a single family unit with 700 sq. ft., built in 1949.
Courtesy of Ed Ohman.

Qué Es Esta? "Eso Es Un Jugete," Dicen Los Niños En Las Escuelas

The Panama Canal Review, October 1, 1954, p. 4.

For the first time Spanish language is being taught in the first three elementary grades of the U. S. schools and school authorities fully expect that it will greatly advance the time when the Canal Zone will become largely a bilingual community.

Recreation Areas For Young And Old Part Of Planning For Zone Townsites

The Panama Canal Review, October 1, 1954, p. 4.

Diablo Heights was the first Canal Zone town for which the townsite planning included small plots of ground located adjacent to new housing areas and developed as open-air recreational areas for children. There are three large playgrounds already established there, and two additional play areas were provided in the planning for the new two-family quarters to replace the 12-family frame buildings. One of the new playgrounds is located on Sibert Street and a second is a half-acre plot between Walker Place and Sibert Street. Both are near the Commissary and can be reached from there without crossing a street. No play equipment has been installed up to now.

Air Conditioned Playhouse

The Panama Canal Review, January 2, 1959, p. 4.

Even the dolls in the Canal Zone are living in air-conditioned comfort these days. A playhouse completed last month at Diablo Heights is not only portable and equipped with running water, but also has air-conditioning.

The dolls with the air-conditioned house belong to Sandra Chesson, 6-year-old daughter of R. W. Chesson, of the Police Division, who built the playhouse during his spare time with the help of his 11-year-old son, Pete.

The telephone, which will be installed later, is an old Army field set and will connect the playhouse with the Chessons' home. Since the playhouse is lined with acoustical tile and is soundproof, this will be an added convenience for Mrs. Chesson in staying in touch with Sandra.

Convenience store opens in old Diablo Clubhouse

The Panama Canal Spillway, July 19, 1985, p. 1.

A 24-hour convenience store opened to the general public on July 3 in the Diablo Clubhouse. "As a kid in school I used to eat lunch here," said owner Liberto Fillo Haro, "and I want to offer the same service to others."

The establishment offers a deli serving a full line of cold cuts and cheeses, typical Panamanian foods like "ceviche," "arroz con pollo" and "sancocho" for take-out, as well as a variety of groceries.

Theatre Guild celebrates 35th birthday tomorrow

The Panama Canal Spillway, December 20, 1985, p. 6.

It will be 35 years tomorrow, December 21, since the Theatre Guild staged its first performance at the Diablo Heights Theater. The plays presented on that date were "Ways and Means," "George" and the "Valiant"; and three of the actors who performed in them—Lollie Levy, Ralph Lindo and Roy Glickenhaus—will be honored tomorrow night, along with others, by the Panama Canal Commission in a special commemorative program scheduled to begin at 8 p.m.

Due honors will also be extended to Stanley Fidanque, the only surviving founding member of the guild still residing in Panama; Star & Herald social editor Anona Kirkland, for her support of Theatre Guild activities over the years; Bruce Quinn, for his efforts in promoting theatrical activities both in schools and in the community; and to others who have made notable contributions.

Long before the current group came into existence, another "Theatre Guild" was established during the Gold Rush by the wives of various Panama City merchants to entertain visitors from the United States who sometimes had to stay in the city for a couple of months awaiting a ship to California.

Following its debut and other presentations at the Diablo Heights Theater, the current guild later moved to a temporary workshop at the former Diablo Heights Dispensary for the production of "Harvey." Then came June 16, 1953, opening night of the original musical "Pretty Lady," the first performance at the current playhouse.

In its 35-year history, the Theatre Guild has presented 182 different productions. It now has 300 members and is seeking even more in an anniversary membership drive.

Memories

Diablo: A Kid's Paradise

Life in Diablo in the 1960s presented an astonishing array of choices for fun and innocent mischief. We could pick cashew apples or mangos for a quick snack, pelt each other with ginnups, play kickball in the sunken playground, or scrounge for soda bottles to sell at the ice dock of the Diablo Clubhouse for .02 each. With our earnings, we could plunk down for a movie at the Diablo Theater, where bats routinely flew in front of the screen, or splurge on a butter brickle ice cream cone.

We played until the streetlights came on or somebody's mother yelled for us to go home. Our adventures began with who had the biggest cardboard box we could flatten and slide down the Clubhouse hill, what treasures we could find in the trash, or whose Christmas trees could we swipe from the heaps of dried-out, icicled trees that dotted the curbs after the holidays. The highlight of the year was the many "tree burns" that various groups sponsored. Who would have dreamed that a future generation of successful adults would have gained their competitive starts by grabbing dried out trees twice their size, running toward flaming infernos at top speed, and flinging the trees into the hellish blaze while barely escaping with their lives? Top it off with riding our bikes behind the mosquito spray trucks and begging for chunks of ice from the milk delivery man you'll have to agree that childhood in Diablo was gangbusters.
- *Melissa Ward Forney*

The End of World War II in Diablo

During the war years in the Canal Zone there was a great deal of sports camaraderie between the military community and the Pan-Canal civilian community. I was at a softball game being played at the Diablo baseball field near the gym between a Navy team and a team from the Zone. There was a big crowd watching the game. I was seven years at the time and lived a short distance from the field in one of the wooden four-family houses on Davis Street. I can't remember why, but I left the game and went home. When I walked into my house, my mother was there listening to the radio. An announcement was just coming over the radio that Japan had surrendered, and the war was over. I ran as fast as I could back to the game and started yelling to people in the crowd that the war was over. No one believed me. Was I relieved when the air raid sirens went off, and an announcement came over the loud speaker that Japan had surrendered. I will never forget the noise and the joy that erupted from the crowd.
- *Jim DesLondes*

Climbing Mt. Everest in Diablo in the Early 1960's

In the early 1960's my family lived in Diablo near the old wooden Diablo Junior High School building. Across the street from the school was a stairway that went up a steep hill that led to the Diablo Elementary School. Right by the stairway was a rocky formation that went up the hill that we kids called "Mt. Everest."

One day I decided to climb Mt. Everest! I got up about four feet high, holding on to the craggy pieces of rock. In a couple of minutes I was nearing the top, quite a feat for a 12 year old! I felt as though I was "Queen of the Hill" until a Canal Zone police car pulled up, and a policeman stepped out.

"Come down off of that cliff!" he ordered.

Heart thumping, I tried to justify my adventure. "I'm doing OK," I assured him. "I'm not going to fall."

"Come down now!"

Humiliated, I had to climb down. When I was within reach, he helped me back down to the street. "Now don't climb up there again--you might fall and get hurt," said the policeman. After he drove away I hurried home, red-faced with shame that I had been in trouble with the law at such an early age!
- *Deborah Ward, former nurse at Gorgas Hospital*

Diablo in the Late 70's

My sister and I grew up as fourth generation Zonians in Diablo, the best town in which to be raised in the Zone. I worked a couple of summers as a student assistant at the Gamboa Gym. Feeling the laid back atmosphere and watching those kids jump off the top of the trestle bridge into the lake, I'll give Gamboa a close second. Balboa and La Boca last; they were always the rivals.

Back to Diablo, my back yard on Sibert Street was the edge of a hill. At the bottom was a giant Banyan tree rising taller than our house on the top of the hill. The tree was our fort and the hill was our giant slip and slide when it rained.

It seemed like we played kickball or battle ball ever day at the Diablo Elementary gym until they kicked us out at closing time. If it rained hard, we played football in the mud until the street lights came on. As we got older and could stay out past dark it seemed like the whole neighborhood knew to go to the school at night to play Ring-a-Levio on the weekends. There would be 20 kids running all over the rooftop of Diablo Elementary; the swamps behind the school were out of bounds. The swamp behind my second house on Morrison Street became our hunting grounds for shooting craps and fish in the ditch at low tide. Later the swamps became our personal motor-cross track during the dry season.

The annual Christmas tree burn is where Diablo ruled. Our parents would help us spend whole weekends going from town to town picking up Christmas trees. Stealing trees from La Boca's pile was always the most fun. The weekend before the big burn we would actually set up camp inside our trees making sure La Boca, Cardenas, and other towns didn't steal them back. With my good friends and family, Diablo had to be the best town, not just in the Zone, but in the whole world in which to grow up.
- *Gregory Gramlich*

Downsizing in Diablo

Beginning in 1984, U.S. school and medical personnel in the Canal Area had to move out of Panama Canal Commission housing into housing that had once belonged to the Panama Canal Company, was turned over to Panama by terms of the treaty, and then was leased back to the military. The luck of the draw resulted in my being assigned to a one-bedroom apartment in an old wooden four-family house in Diablo. At first the location--within walking distance of Diablo Elementary School, where I was working —seemed like the only good thing about the forced move. Like just about everyone else in our area, I was coping with the strain of having to downsize and such inconveniences as having to spray frequently for cockroaches and having a hot water heater break down. Eventually, though, many of us began to see a few silver linings and laughs in our situation--living closer to friends that had been scattered in various areas before, getting bigger and better hot water heaters when the old ones broke down, living across the street from the Home for Unwed Mothers, etc. When the curtains of my living room were arranged in a certain way, the only thing visible outside from inside was a big, gorgeous tree right there next to the house. I loved it and managed to (sort of!) convince myself at times that I was living in a mighty fancy tree house with at least some of the comforts of home!
- *Kathy Egolf*

What I Won't Tell You About Diablo

I won't tell you where the best mango trees were or how my brother would sell duros for a nickel or limes he picked for a penny or about riding bikes to the Spinning Club or playing ring-a-leeveo in the street at night or going to the Owl Show and watching bats fly in front of the screen. I also won't tell you about our famous Christmas tree burns (or stealing trees to make ours the best!). I will tell you that when my family moved, it was to the other side - of our duplex - that's as far as we were going from our home in Diablo.
- *Valerie K. Shaw, BHS Class of 1980*

Outside our house at 5184-B Parsons Street in 1969. Then we moved to the "A" side. *Courtesy of Ruth Shaw.*

TIME IS FLYING!
We Are All Counting
The Days to Christmas!

★ ★

Shopping will be a pleasure if you go to your favorite Clubhouse Record Section and browse through our large selection of phonograph records. We have a wide choice of Christmas records, as well as records for children and adults — from bedtime stories to swing and symphony. We also have record storage albums that would make very suitable gifts. Your Christmas shopping problems will be solved by giving phonograph records — a gift that will give lasting pleasure throughout the year.

CHRISTMAS ALBUMS AVAILABLE
78, 45, 33-1 3 rpm

"Carols for Christmas"
"Merry Christmas" Bing Crosby
"Under the Christmas Tree"
  by Jan Garber's Orchestra
"Christmas Greetings" Bing Crosby
"Carols" by St. Luke's Choristers
  Ethel Smith, "Christmas Music"
"Night Before Christmas"
  by Fibber McGee
"Twas the Night Before Christmas"
  by Fred Waring's Orchestra
"A Visit from St. Nicholas"
"Christmas Cheer" Andrews Sisters
"Music for Christmastide"
  by the Rome-Vatican Choir
"Christmas Favorites" by Vic Damone
  and Frankie Lane
"Christmas Music and Stories"
  by Two-Ton Baker

SINGLE RECORDS AVAILABLE
78 and 45 rpm

"Christmas Carols by the Old Corral,"
  Tex Ritter
"Jingle Bells" J. Mercer
"Little Town of Bethlehem"
"Frosty the Snowman"
"Rudolph the Rednosed Reindeer"
  Crosby "Christmas"
"Silent Night" Fred Waring
"The First Nowell" Fred Waring
"White Christmas" by Crosby, Cavallaro
  and other artists
"Santa Claus is coming to Town"
  Bing Crosby
"Joy to the World"
"Hark, the Herald Angels Sing"
"Santa's Toy Shop"
"Night Before Christmas"
  by Arthur Godfrey
"Hymns for Children"
  Brahms "Lullaby"
"Mother Goose Songs"

AND HUNDREDS OF OTHERS BY ALL POPULAR ARTISTS!

PRICES RANGE FROM 25c. UPWARD

WHERE ELSE CAN YOU GET SO MUCH PLEASURE FOR SO LITTLE?

★ ★

Christmas Greetings from Panama

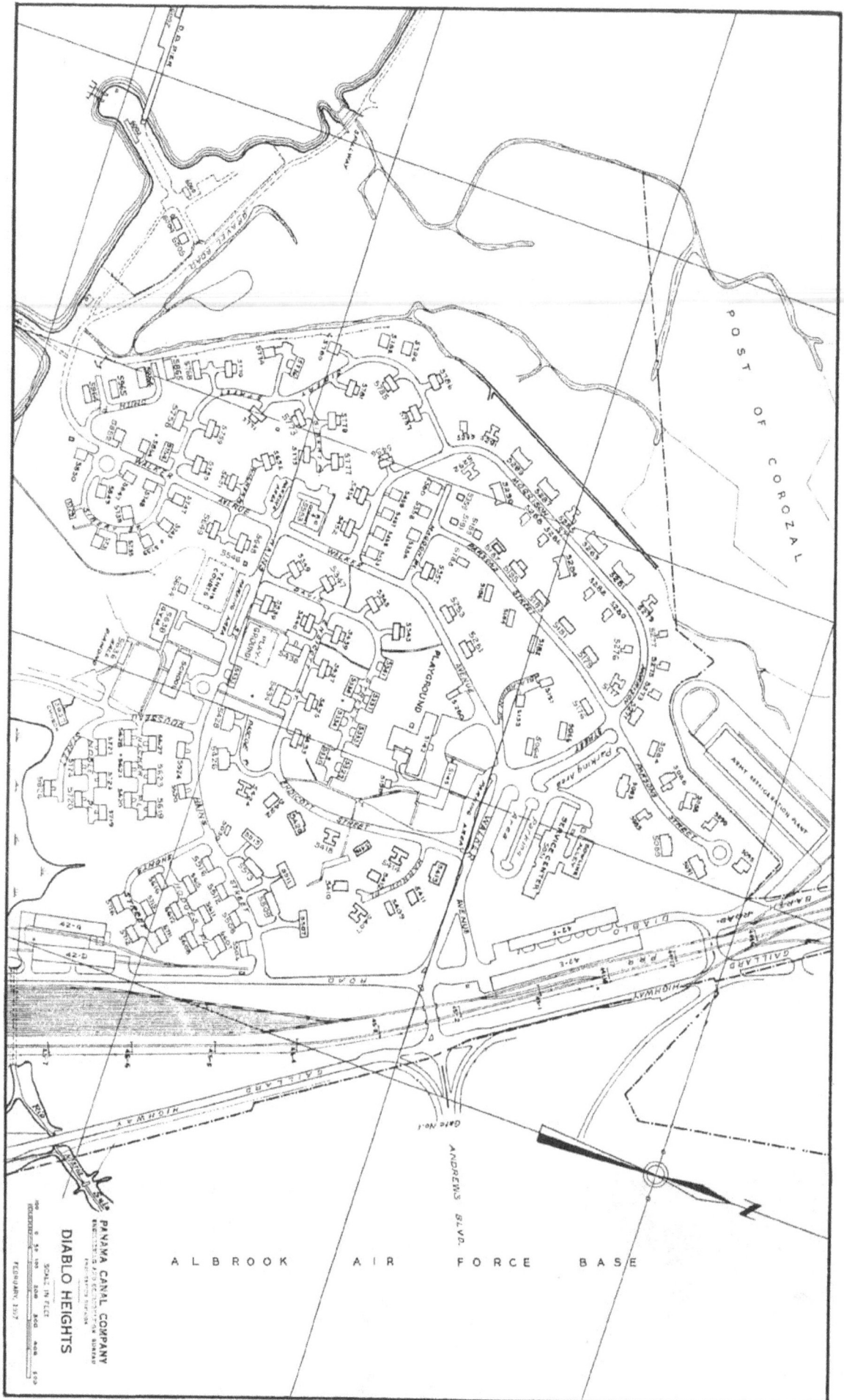

PANAMA CANAL COMPANY
ENGINEERING CONSTRUCTION
ENGINEERING BUREAU

DIABLO HEIGHTS

FEBRUARY 1957

POST OF COROZAL

ALBROOK AIR FORCE BASE

Gamboa

Sponsors

Baglien Kids: Lynn, Beth, Julie, & Joel
Stella Bradney
Harry, Mary, Katherine & Billy Egolf
Leon & Ruth Egolf
Andy Nash English
Peggy (Hale) Huff
Charles W. "Chuck" Hummer
Barbara Bartholomew Krueger
Bob & Cheryl Russell
Val & Bert Schroeter Family (Suzie & Tina)
Michael & Elaine Stephenson Family
The Lynn Stratford Family
The Wards—Marvin, Jackie, Debbie, & Lissa

Your Town - Gamboa

The Panama Canal Review, September 4, 1953, pp. 8-10.

TWO MIGHTY CRANES which tower above their surroundings are outstanding landmarks for Gamboa. They were featured several years ago in a moving picture of the Canal Zone. Built in Germany, the two cranes— *Ajax* and *Hercules*—were towed across the Atlantic and their superstructures added here in 1914.

Until Gamboa became the headquarters of the Dredging Division in the fall of 1936, it had, as a town, played no important part in Isthmian history, either during the colonial period or the time of the buccaneers. It was not even a railroad stop until about 1911.

Today it is one of the most attractive communities in the Canal Zone. From its docks, dredges and cranes and barges and tugs go about their business of keeping the Canal open for traffic. Its community buildings are grouped within a fairly small area and its quarters-lined streets run up into hills which during the late dry season blaze with the brilliant yellow guayacan or the pink and purple of other flowering trees.

There was no Gamboa during the day of the Spanish colonists. In its approximate location was a small river town called Santa Cruz which historians believe may have been a place for discharging boats during low stages of the river. Three miles up the Chagres was Las Cruces where trans-Isthmian travelers of those days changed from boats to burros on their way to Panama City.

When the Panama Railroad was built in 1855 its route followed the west bank of the Chagres through Matachin and Gorgona, nearly opposite present Gamboa, to the river bridge at Barbacoas, 16 miles south of Gatun. No town at the present location of Gamboa was shown on maps of those days.

French Dam

As a construction point for Canal work Gamboa (which means a tree of the quince family) first came into prominence when the French Canal Company began excavation.

French plans for a sea level canal called for a dam across the Chagres River at Gamboa to retain the Chagres in a large lake while a channel known as the east diversion, carried its waters to the Atlantic.

In 1887 when the French Company switched to a temporary lock canal, they continued planning for a Gamboa dam. This would have supplied water for the locks which were to be built at Bohio Soldado about 8 miles south of Gatun on the Atlantic side and between La Boca and Paraiso on the Pacific side.

Over the Chagres at Gamboa the French built a bridge over which materials were hauled across the river and to the nearby spillway. The bridge was about 365 feet long, the north span being a girder about 58 feet long. In a flood in 1890 this girder was carried away and the pier on which the channel end of it rested was tipped. When work on the Panama Railroad relocation bridge at Gamboa was started in 1907, the pier was righted and the two truss spans used for construction purposes.

Chagres Dammed

Flood control for the Chagres, now provided for by Madden Dam, was an early concern of the American Canal forces when they took over the Canal rights in 1904. A large field party was sent to look into the possibility of building a dam at Gamboa. The idea of this dam was abandoned when the lock type canal was decided upon, and plans were made to form Gatun Lake by damming the Chagres at Gatun.

With a lock type canal some provision had to be made to prevent Culebra, now Gaillard, Cut from flooding from freshets in the Chagres River. In 1908 an earth dike was built across the northern end of the Cut, approximately opposite the present location of the penitentiary. During the 1906 flood, the river had risen to 81.6 feet at Gamboa, but this was before the dike was built and before the Bas Obispo (about 10 miles north of Pedro Miguel) section of the Cut was completed.

Railroad tracks ran across the top of the dike, originally 73 feet above sea level. When Gatun Lake began to fill, in 1912, the top of the dike was raised and strengthened.

Dike Dynamited

On October 10, 1913, the dike was blown up and the lake water permitted to rush into the partly filled Cut. Details of the dynamiting have been told many times: How President Woodrow Wilson, in his White House office, depressed a lever, relaying electric current over land telegraph to Galveston and submarine cable across the Gulf of Mexico and the Caribbean to trip a weight attached to the handle of a switch in the Canal Zone. The weight threw the switch and set off the blast. Half an hour or so after the dike was broken a cayuco made the passage through, followed by three launches.

By this time, a few buildings were beginning to appear in Gamboa, where there had been little except the old bridge and a few houses left by the French which, presumably, had been occupied by their hydrographic forces and the dam workers. A 1906 report mentions repairs made to a fluviograph tower and 19 buildings used as quarters.

Gamboa Townsite

In July, 1911, a Canal Commission committee recommended that a townsite of Gamboa be located on the first dump north of the Gamboa bridge, to house people in the seven mile stretch between Tabernilla and Gorgona who would have to move as the lake waters rose. At this time some 700 people lived in what is now Gamboa. Of these 203 were "Silver" roll employees, the remainder probably their families. No Americans are shown on the census of that date.

In 1914 Gamboa's population was down to 173. There was a police station, a four-family house which had been brought from Empire, a two-family house from Culebra. Bachelors lived in old box cars and south of the bridge more box cars housed married and bachelor "Silver" employees from the gravel plant. A married American employee had quarters in an old railroad tower. The commissary was made up of three box cars, according to Emmett Zemer, now of the Community Services Bureau, who was Gamboa commissary manager, and the town's fireman, for eight months in 1916. The commissary's main function was to provide provender for Dredging Division personnel working in the nearby

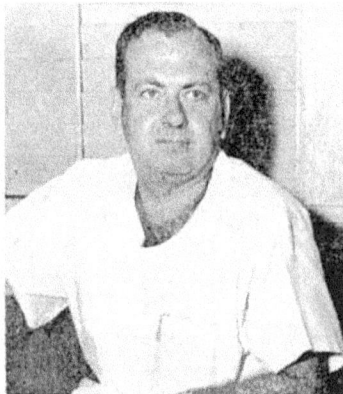

DR. ALBERT BLANSHAFT,
Gamboa's Doctor

MRS. ELSIE HAUGHTON,
School Principal

SGT. HERBERT HOLMER, Police Station
Commander

C. B. PRESCOTT, Manager, Santa Cruz
Clubhouse

P. B. HUTCHINGS, Housing Manager

WILLIAM K. McCUE, Postmaster

section of the Canal. Each day anywhere from one to three box cars of cold storage and staples were pulled into the spur to be unloaded before dawn and reloaded into launches which ferried supplies to the floating equipment.

Two other Gamboans of those days were A. C. Medinger, just retired as Railroad and Terminals Director, and J. A. Fraser, now dredging supervisor for the division. Then young bachelor engineers, they lived in one of the four-family apartments.

Mosquitoes And Malaria

The mosquitoes were terrific. Mr. Zemer had malaria twice in his eight Gamboa months. Because of the high malaria rate the Chief Health Officer recommended no more building at "Gamboa Cabin," the section around the railroad station. Instead, he said, future building should be on the south side of the Chagres. This was the beginning of an argument which lasted many years.

Early in 1917 a building sites committee was appointed to look into the advisability of moving what quarters there were in Gamboa across the Chagres River, to the general vicinity of the gravel plant. Col. Jay J. Morrow, later Governor but then Engineer of Maintenance, objected to the change and three months later the committee went along with him. One objection, they said, was that if the railroad station were moved south of the bridge at least two coaches of every train would stop on the bridge, jeopardizing passengers who might get on or off.

During all this time, Dredging Division headquarters were located at Paraiso. That they were ever moved to Gamboa was due largely to an accident and to the persistence of one man, the then Dredging Division Superintendent, John G. Claybourn.

On July 30, 1923, Mr. Claybourn wrote a memorandum to Governor Morrow, recommending that the Dredging Division shops be moved from Paraiso to Gamboa for two reasons: "First, as a safeguard in case of obstruction of the Cut by slides, the logical location being between any possible dredging and the dumps in Gatun Lake; second, increased Canal traffic, as well as the size of ships, introduces a serious menace to our fleet when moored in the comparatively narrow confines of the Cut at Paraiso."

Sunken Submarine

Three months later his first argument was vividly and tragically emphasized. On October 28th the S. S. Abangarez and the submarine O-5 collided in Cristobal harbor. The submarine sank. Working against time to save the men imprisoned in the submarine, salvage men sent out a hurry call for one of the Dredging Division's big cranes.

But just about that time 300,000 cubic yards of rocks from the west bank slid into the Canal prism, leaving a channel only 120 feet wide through which the bulky crane had to be maneuvered. It did get through and did raise the O-5 in time to save the lives of the men inside.

For almost 13 years, Mr. Claybourn urged successive governors to consider the transfer to Gamboa. The north-of-the-bridge and south-of-the-bridge argument was renewed. In 1928 the outgoing Governor, M. L. Walker, passed the problem on to Col. Harry Burgess, soon to succeed him, saying that the 'transfer "would be so expensive that it is futile to consider it at present---I do not consider it advisable to

LT. JOHN YOUART, Fire Station
Command

J. H. SMALL, Clubhouse Luncheonette

ROBERT J. BYRNE, Commissary Manager

A. L. MORGAN, Santa Cruz School Principal

include it in next year's estimates so you will have to wrestle with it later."

Several newspaper accounts of the proposed change appeared in the next few years but it was not until April 1933, that Governor J. L. Schley appointed a three-man board to look into the question of a Gamboa townsite. They reported that the move was feasible and would cost about $2,780,000 spread over a three-year period. That year there were only 251 residents, 10 of them Americans, living in Gamboa.

The transfer was a terrific project. It meant the building of shops and office buildings, schools, a gas station, fire station, commissary, clubhouses, a garage, a dispensary and all of the other buildings necessary for community living. Streets, sewers, power and telephone lines had to be installed; landscaping had to be done.

On January 13, 1936, J. F. Evans, now manager of Balboa Commissary, took over Gamboa's one-room commissary-postoffice combined. He opened the new

commissary some months later and was relieved of his postmaster's duties in April by Gamboa's first feminine postmaster, Gladys A. Houx.

The first Dredging Division families moved to Gamboa in September 1936. From 280 people in June 1936, the population jumped to 1,419 in a year, and 2,132 in June 1938. Gamboa's peak population was 3,853, in 1942.

With their own hands and their own money, the people of Gamboa built the Civic Center, now headquarters of the Civic Council and scene of most community activities. It was first a USO to provide recreation for the thousands of soldiers stationed in the hills nearby. Women of the town worked night after night, arranging dances and parties, providing cake or cookies or pies.

Another community project, also built with Gamboa labor and money was the Gamboa Golf and Country Club in the Ridge section, overlooking the Chagres. The

women brought picnic lunches or tended barbecues while their men hammered and sawed and poured concrete. The club was opened officially on January 1, 1939.

Today Gamboa is somewhat changed. The combined population of Gamboa and the local-rate settlement which has been known since 1948 as Santa Cruz, was 3,353 last June. It now has five churches; its Yacht Club members tie up their boats along the Chagres River docks. The big clubhouse is closed and Henry Grieser, famed swimming coach, no longer supervises the human tadpoles at the pool.

But the people of Gamboa still like it. They like the wider streets, named for Canal officials--Morrow and Goethals Boulevards--or old Dredging Division men--Pratt Parkway, for instance. They think the Gamboa stars shine brighter at night; Gamboa breezes blow cooler; Gamboa grass is greener. Confirmed Gamboans agree with their doctor, Albert Blanshaft: "It's the best town on the Isthmus."

Santa Claus From Gamboa

The Panama Canal Review, January 1, 1954, p. 12.

BOOTED AND FURRED and straight from Gamboa, not the North Pole, Santa Claus visits the Paraiso kindergarten just before Christmas. Later he stops at La Boca and Santo Tomas Hospital. It is not strange that his helpers are in police and police guard uniforms, for Santa has come from Gamboa penitentiary to bring gifts that convicts there made during the year for some of the young folk of the Canal Zone and Panama.

Canal Zone Penitentiary Resembles Santa's Workshop

The Panama Canal Review, December 3, 1954, p. 11.

Convicts at the Canal Zone Penitentiary, who have won the privilege through good behaviour, are busy in the shops sawing, hammering, painting, and otherwise fashioning colorful toys for Christmas distribution to poor and needy children on the Isthmus. Under the supervision of Policeman Karl D. Glass, they are making dolls, wagons, toy animals, jigsaw puzzles, fire engines, and numerous other articles.

The project is part of the penitentiary's rehabilitation program; this is its third year.

Materials for the toys are bought with money from the Convicts Welfare Fund or are received from donations from residents of Panama and the Canal Zone.

Left, Type 201 is an 8-unit apartment complex built in 1938. Each apartment had a 560 sq. ft. living area. There were six built in Gamboa and two in Balboa. Right, Type 210 is a 12-unit apartment complex with 635 sq. ft. of living space in each unit. There were twelve of these complexes located in Gamboa. *Courtesy of Ed Ohman.*

Gamboa Pool To Be Closed

The Panama Canal Spillway, Friday, August 1, 1997, p. 4.

Because of economic measures, the Gamboa pool will close at the end of the fiscal year. In service to the Gamboa community for more than 55 years it provided both recreational and job-related services to Canal employees. The pool was built around 1938, when the Dredging Division was moved from Paraiso to Gamboa. Swimming is an important skill for Canal employees who work on or near the water. More than 4000 Canal positions list "swimming ability" as a job requirement, and pool employees help Canal personnel prepare for the swimming test with special classes. Employees must also retake the swimming test every three to five years to assure they are still qualified swimmers.

There are two types of tests, one easy and one hard. Both tests require that the employee enter the water wearing coveralls and tennis shoes. Those who are occasionally near the water must be able to swim 100 feet and remain floating for three minutes (the "easy" test). Those taking the "hard" test, seamen, floating equipment operators and the like, must swim 200 feet and remain floating for six minutes. In addition to the required tests, employees whose work takes place on the water may also take a "Basic Rescue and Water Safety" course. The course teaches them how to prevent water-related accidents and how to help themselves and others in the case of an emergency.

A "Summit Meeting"

The Panama Canal Review, March 1964, p. 8.

Summit Gardens is a center of recreation for Canal Zone area residents. Dancing, picnicking, playing games, and barbecuing are the four items that take up most of the time on a weekend at this delightful spot. There are swings and a merry-go-round as well as a zoo full of interesting animals.

Memories

The HYACINTH II

There is a significant amount of aquatic vegetation on the Chagres River in the form of floating grass and "hyacinth," a tough plant with big leaves above the water and strong roots beneath. The grass & the hyacinth can form large floating islands that drift downriver toward Gamboa and the Canal. These islands can block water intakes for the Canal Zone's water system, and they can foul vessel's propellers and engine cooling water intakes.

The Dredging Division had a crew of men (Hyacinth-men) whose job it was to remove these islands. They worked from six or eight 14 foot wooden rowboats (pangas),two men to a boat. One man rowed the boat; the other used a pitchfork and machete to load debris into the boat. Then they rowed over to shore and pitched it onto the bank. Hard work. All day under the sun. Rain or shine - and it can really rain in Panama.

The tender for this fleet was a launch named the HYACINTH II. She was an odd looking craft, low to the water (so she could get under the Gamboa Bridge), solid steel, heavy, heavy, about 30 feet long. Originally, she had a steam engine; but when I knew her, in the late 50's & 60's, she was diesel.

Every weekday morning at 0700, the men piled into their pangas and formed a line with each panga tied to the stern of the other. The HYACINTH II then towed this fleet to the day's work location. They looked like a flock of ducks following their mother.

I got to know these men because I would paddle my cayuco on the Chagres and around Gamboa and up the Mandinga River. The crew was friendly. The boat had a big water cooler and those paper cups. I'd paddle alongside to get a drink and some shade. While the crew worked, the bow of the boat would be stuck in the bank of the river. The boat's operator was a heavy set man, jovial. The foreman, also heavy set, wore a wide brimmed hat - looked like a plantation boss.

The old boat - last I saw her - was floating at Gamboa boat club. Abandoned, neglected, forgotten by most, but not by me. Good memories, good people.
-Gerry Cooper

A Scary Night in Gamboa

Many of us have raved about what a wonderful place the Canal Zone was in which to grow up and raise a family. And those who lived in Gamboa always seem to be especially strong in singing the praises of that very special small town halfway across the isthmus. But there was one night, perhaps in 1950, when my family was only too painfully aware of the many possible dangers that surrounded us there. Early in the evening we discovered my younger brother, Billy, three or four at the time, was nowhere to be found in his usual haunts near our house at 155-A Williamson Avenue—perhaps out wandering after chasing the DDT truck. While my father and the ever helpful Canal Zone Police looked for him, all of us worried about the risks the nearby canal, railroad tracks, and jungle posed for such a little boy on his own at night. After what seemed like an interminable amount of time, but may have been only a couple of hours, Billy was finally found--the only white face in the choir at the Catholic Church in Santa Cruz, about a half mile from our house. It was a joyful reunion on the part of everyone--Billy, the choir members, the priest, the police, and our family! For years afterwards, a number of the Santa Cruz residents would ask us how Billy was whenever they saw a member of our family.

- Kathy Egolf

DDT Fogging Trucks

The Trail of the Tarantulas

When I was three or four years old, my family lived in house 155-A on Williamson Avenue in Gamboa. The two-bedroom duplex was off the ground with a basement and one-car garage. Most of the houses in Gamboa had dense vegetation surrounding them – huge vines hung from the trees and all kinds of critters were always crawling out of the canopy. My brother and I played with the little spiders and insects that lived with us in both the house and the basement. In the garden, we got a kick out pushing dirt into the little doodle-bug craters and watching them kick themselves out.

My father was a fireman and worked twenty-four on, twenty-four off. I remember the stacked porcelain containers that my mother would put his dinner in when he had to spend the night at the fire station. Off we'd go in the car down to the fire station to visit with him while he ate. One night when we got back to the house it was dark, and as my mother maneuvered the car into the garage, we all screamed in horror. The car headlights lit up the garage walls and they were covered with the largest spiders I'd ever seen – thousands of them. Well, probably only a few - but at the time it seemed like thousands. We couldn't roll the windows up fast enough as she backed the car out; we left it at the curb overnight. I laid awake all night scanning my bedroom walls for fear that those hairy monsters had followed the little spiders into the house and would be crawling all over my bedroom – and me.

It wasn't too long after the garage incident that we went to the old Gamboa Theater one Friday night to a movie. There were a bunch of us little kids; our mothers parked us all in the front row and they went up in the back. In the opening scenes of the movie I knew I was in deep trouble. There was a mean looking old scientist with frizzy hair and thick glasses. He had cages full of large, hairy spiders, just like the ones that were in our garage. Well, one of them got out of his cage and escaped into the night. At this point I had become part of the seat, terrified of putting my feet on the floor and trying to find my mother in the dark recesses of the theater. All of us in the front row had a death grip on one another. Minute by minute the spider got bigger and bigger until he could straddle an entire house, one that looked about the same size as Gamboa Theater. How I ever slept another night in Gamboa I'll never know – spiders in the house, under the house, maybe over the house. I lived in mortal fear. I never again sat in the front row of Gamboa Theater.

- Peggy Huff

Moving . . . Again?

Moving to/from town sites was a way of life in the Canal Zone. As with most new Canal employees, our first home was a 4-family wooden house on the Ridge in Gamboa. Due to the 40 minute-drive to work in Balboa, we soon moved to the similar type house in Diablo. It had the same thin walls and "see through" wooden floors. It was through those floors that stereo equipment belonging to our downstairs neighbor was christened....by our dog.

A few more years of service got us into a wooden duplex in Gavilan with not so many neighbors to share walls/floors with but about the same number of roaches and termites that seemed to reside in all wooden houses. That came to a happy end when we moved into new 2-story concrete quarters in Williamson Place.

We thought that our next move to a cottage in Los Rios would be our last until retirement, but one more took place...while Bob was in the States. It was an easy move for me, with the help of friends and no questions asked regarding hanging pictures or arranging furniture. We teased about leaving Bob a note at the Los Rios house as to the move to a 2-story duplex in Cardenas, but decided to forewarn him before he got back from his trip! Packing out of that house to move to the U.S., we thought of the saying, "If these walls (of our houses) could speak, the stories they would tell."

- Bob and Cheryl Russell

Gamboa Stories

My parents, Herbert Holmer and Mary Joyner, were married in 1938 and initially lived in Pedro Miguel. From there they moved to Barnaby Street (Balboa flats) for a couple of years, then to Ancon (at the foot of Gorgas Hospital). When we moved to Cocoli, my dad was the station commander for the Canal Zone Police Division. We lived south of the Police Station and west of the elementary school, sharing a duplex with the William Baldwin family. From Cocoli we moved to Gamboa and lived at 146 Harding Avenue until moving to Buffalo, N.Y. in 1953.

I learned to swim (and swam competitively), thanks to our neighbor Mr. Connors, who took the time to teach me individually in spite of his busy schedule. One day I was walking around in the 4-foot pool and stepped on something sharp. I thought it was a piece of glass but after hurrying to get out of the pool, I saw an iguana rising up from the bottom of the pool.   During mango season, I spent a lot of time climbing trees and eating tree-ripened mangoes to my heart's content. I ate a lot of rose apples as well. Two of my greatest challenges while living in Gamboa were riding my bicycle up the steep hill to our house and

Gamboa Elementary School.
Courtesy of Peggy Huff.

also riding it across the one-lane bridge before the traffic light changed.

My father made mahogany louvers for all of our windows on the front side of our house. He also grew hundreds of African violets which he gave or sold to the Panama Canal Company for a garden/lawn party they were having during Queen Elizabeth's visit.

To grow up in the Canal Zone, as I did, was to grow up in a tropical paradise.
- *Veronica L. Homer Mahaffy*

- -

When I was a small boy, about the only thing to do was to play in the jungle. We used to build forts and go hunting for lizards with our B-B guns. We would just naturally goof around just like any other kid used to do.
One of the best places to go and hide out when we didn't want to go to school was the waterfall just before the Gamboa bridge. One of the main events in Gamboa came around Christmas time. We used to try and collect all the Christmas tree we could find, and then we would make teams to see who could collect or rip the most trees from each other. When the big night came, we used to burn the trees at the Civic Center and have a wienie roast.

The funniest thing that happened at school was when my sixth grade teacher wasn't watching what he was doing, turned around, ran into the chalkboard and broke his nose.

One of my favorite tricks was to take some scrubber on a snipe hunt. We used to take a kid who didn't know what a snipe was, take him way up on a hill and give him a paper bag filled with rocks and sticks. We used to tell him to sit down, hide in the bushes and rattle his bag, and the snipe would come running to us. So we would all go back home and wait for him to come back. Well, when it got dark and he still hadn't come home, we would go and get him—that is, if he hadn't already found out that there really isn't any such thing as a snipe. We would get a big laugh out of that.

Two of my best friends and I had nicknames. Their two names were Magpie and Gumbie. Magpie talked so much we could never shut him up. And Gumbie was dumb and dorky. Mine were Fred and Zote. I can't remember why they called me that.
- *Ferron Coombs in 1978 for a folklore class taught by Mary Knapp at Balboa High School.*

- -

In Gamboa we used to have all different kinds of games like dare games to join a certain gang. There was a big gutter in the back of our house, pretty deep, real wide, and it had three poles going across. You had to take your bike down the hill in the jungle and cross the bar on your bike. If you couldn't do it, you couldn't join the gang.

There was a hill in Gamboa that had water tanks on top. There was a story that there was a black panther up there. They would burn stuff back there; and when the fire was up, they said the panther would be out looking for somebody.

When I was very small, about three or four years old, I had this red tricycle. A goup of us used to race down this hill called Black Hill or Black Road. We wouldn't even have to pedal. The last guy had to do something; I don't remember what and if you didn't do it, we would get to do something like tie the kid up and leave him there till his parents came for him.
- *Rick Doubleday in 1978 for a folklore class taught by Mary Knapp at Balboa High School.*

Gamboa Sea Scouts

In 1954 the SeaScouts of Gamboa invited some of the girls of Gamboa to go for the day to Taboga. We went in the two Sea Scout Boats. On the way home the small boat had problems so we had to transfer some of the kids on it to the big boat. It was a bit choppy and some of us started feeling a bit queasy, but I don't remember anyone really getting sick. On arrival at the Yacht Club we were given a small shot of brandy, and then we "posed" for the picture.
- *Doris Monaco*

Back row from left: Camille Ellis, Steve Herring, Fred and Elaine Saunders (chaperones), Olga Holmes (another chaperone), a friend of her husband's.
Middle row: Butch Hope, Donna Idol, Doris Ehrman, Jackie Dunn, Rex Daisey, Henry Ehrman, Gerry Simon, Laura Dew, Dick Gramlich.
Front row: Herbie Spector, Don Ryter, and Jim Richardson.

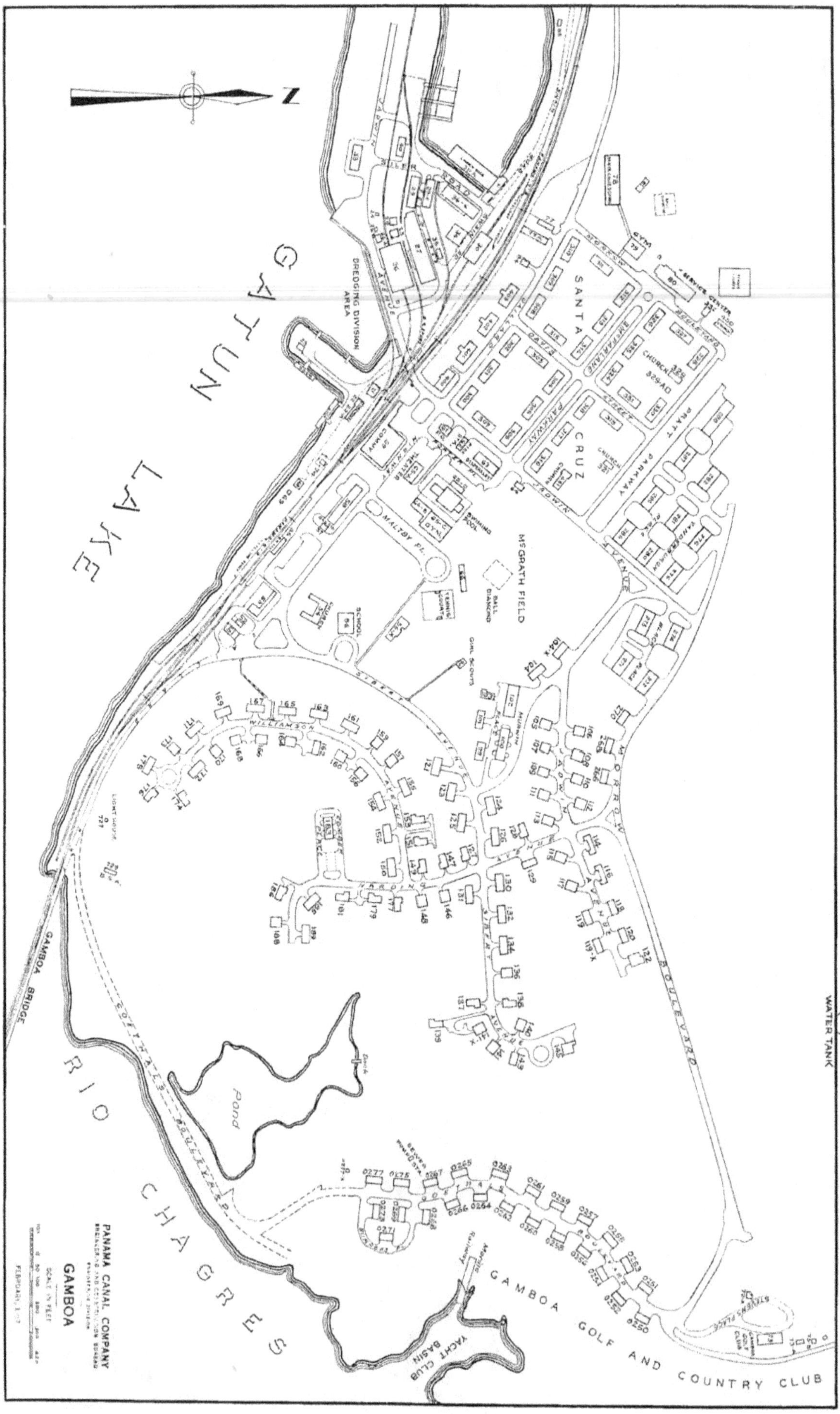

PANAMA CANAL COMPANY
ENGINEERING AND CONSTRUCTION BUREAU
ENGINEERING DIVISION

GAMBOA

SCALE IN FEET

FEBRUARY, 1973

Gatun

Sponsors

Caleb, Ruth, Orrin, Mary, Cub & Alice Clement
The Joseph & Carol Coffin Family
Louie, Barbara & Jon Dedeaux
Leon, Ruth, Harry, Ruth, Dick, George & Barbara Egolf
Andy Nash English
Dennis & Peggy Huff
Ann Thomas O'Neal
Elma L., Ernestine & Fred Raines
Virginia Kleefkens Rankin
Michael & Elaine Stephenson Family
The Lynn Stratford Family
Charles E. Thomas
Anna Dorn Thomas
Charles A. Thomas
J. E. Dorn Thomas
Family of James & Stacia Walsh

Your Town - Gatun

The Panama Canal Review, November 6, 1953, pp. 10-12.

GATUN'S TRIPLE FLIGHT of locks, extending one and one-sixth miles, raises or lowers transiting ships 85 feet. Islands which were once hilltops dot the surface of the lake beyond the upper end of the locks.

Anyone revisiting the Canal Zone today, after an absence of 40 years, would have considerable trouble orienting himself in the town of Gatun. Its topography has been more changed and the town itself has undergone more metamorphoses than almost any other section of the Canal Zone.

The name El Gatún, for village and river, appears on maps of Panama's colonial days. It may be derived from "gato," for cat, referring to the feline smooth-running river; or it may come from "gatunero" or seller of smuggled meat, since Gatun was known as a place where stolen cattle were brought for sale to travelers.

Sir Henry Morgan and his men bivouacked close to Gatun, near what is now known as Navy Island, after sacking the old city of Panama nearly 300 years ago.

During colonial times and until the beginning of this century, Gatun was located on the west bank of the Chagres, about where the office and machine buildings of Gatun Dam now stand. In the mid-1800's it was described as a sleepy village of 40 or 50 cane huts, on the edge of a broad savannah. On a hill overlooking the river were ruins of an old Spanish fort.

The gold rush of 1849 and the beginning of construction of the railroad a year later woke Gatun with a jolt. Travelers, on their way upriver from Chagres, paid 25 cents each for eggs and $2 a night for a hammock, exorbitant prices for those days.

When work began on the railroad, ships carried machinery, provisions, and part of the railroad force up the Chagres to Gatun. From Gatun they worked their way back through the swamp toward the railroad's Atlantic terminus on Manzanillo Island, now Cristobal-Colon.

A month after the railroad ran its first work train, on October 1, 1851, as far as Gatun, a "norther" forced two passenger-jammed ships into Limon Bay. The thousand California-bound gold hunters, unable to land at Chagres and start their journey up river from there, demanded passage on the railroad. They paid 50 cents a mile and $3 per 100 pounds of baggage for the 7-mile train ride.

As the railroad tracks stretched further toward the Pacific, Gatun became just a railroad station and a river produce landing. Beside the tracks which ran on the east bank of the Chagres were a large, two-story house, a cluster of smaller buildings, and "suitable outbuildings" around a flourishing garden.

But about 1880 the French Canal Company forces reached Panama. Almost overnight, thousands of prefabricated buildings were unloaded from ship after ship. Warehouses, quarters, and machine shops went up in Gatun and along the railroad line. By 1881 Gatun, rechristened Cité de Lesseps, had become the largest town in what is now the Canal Zone.

After the French virtually abandoned work on the canal, Gatun lapsed into the quiet of its pre-boom days. American forces began work in 1904 but Congress did not authorize a lock-type canal until 1906.

French engineers and the first U. S. Isthmian Canal Commission had planned to dam the Chagres at Bohio, about 17 miles from Colon. It was John F. Stevens, the Canal's second Chief Engineer, who advocated harnessing the Chagres at Gatun.

"Why not make the Chagres the servant instead of the master of the situation?" he asked.

Engineers quarreled with his selection of Gatun as the dam and lock site and declared that the rock foundation was not suitable. Stevens held firm, and declared: "If Nature had intended triple locks there she could not have arranged matters better." But it was not until the then Secretary of War, William H. Taft, brought a group of engineers to the Canal Zone—they pronounced the location satisfactory—that the furor died down and work could be started.

Tent City

While the family and bachelor quarters and labor barracks to house the lock and dam forces were being built, the workers and some of their families were sheltered in about 150 tents of varied shapes and sizes which stood in more or less orderly rows alongside the railroad tracks. The Labor and Quarters Department objected roundly.

Jackson Smith, its head, predicted: "On account of its being a tent city, the men will not remain there after their first pay day;" and his assistant, Lt. R. E. Wood, now Chairman of the Board of Sears, Roebuck, added: "Gatun is going to be what Mount Hope and Comacho have proven to be—a sinkhole for men."

THE LESTER L. LARGENTS are a Gatun-man-and-wife team. Seargent Largent is in charge of the Gatun police station; Mrs. Largent is the nurse in charge of the Gatun first aid station.

MISS RUTH CROZIER, Gatun School Principal

The town was built under difficulties. Before any houses could go up, a 16-foot plank road had to be laid from the railroad tracks to the foot of a steep hill and all material had to be carted over the road and up the hill. Despite the difficulties, 97 buildings had been erected by June 1907 and work had started on a commissary in a hollow opposite the present police station.

Sibert's Hill

A year later Lt. Col. William L. Sibert established the headquarters of the Atlantic Division at Gatun. The building was on high land just north of the present railroad station and close to the bridge over which Bolivar Highway now crosses the tracks. A metal hitching post to which Colonel Sibert tied his horse is still in place there, a metal plaque in its base.

From the porch of the wind-swept office building, old-timers recall, there was a splendid view of Limon Bay and the harbors of Cristobal and Colon, the dredges at work in the approach channel, the locks under construction in the valley below and beyond them the dam which was beginning to take shape.

For his residence, Colonel Sibert chose a hilltop east of the village on the road then being built from Gatun to Cristobal. Subsequent revampings of the town have leveled it off.

In the meantime rapid progress on the locks and dam meant that the railroad, which ran close beside the Chagres, had to be relocated on spoil taken out years before by the French. The river had already been diverted.

Moved To New Town

In April 1908, the old native village and its 600 inhabitants were moved to "New Town," which was located just about where the third locks excavation was dug over 30 years later. As it was rebuilt New Town had

MISS PEARLINE CARTER, Chagres School Principal

over 110 buildings including a church and its parsonage, and about 25 stores.

Gatun was beginning to assume the look of a town. The railroad was moved from what is now the west side of the locks to its present location; the present station was begun in 1909. The same year work was started on a new two-story commissary at track level, north of the railroad station with the entrance at the bridge level. In 1909 a $25,000 clubhouse was built on a knoll next to the present dispensary.

There were schools, a two-story hotel—its front lawn bore the letters "Q. M. D." (for Quartermaster Department) in foliage plants—a post office and telephone exchange near the present intersection of Bolivar Highway and San Lorenzo Street. There was a two-story lodge hall, which also served as a church, opposite the present dispensary, and bachelor quarters, one of which was located where Sibert Lodge now stands. The dispensary was on the location of the present Gatun school.

A row of big quarters—which housed the families of such people as Maj. Chester Harding, who was in charge of locks construction and was later the Canal's second governor, William Gerig, who headed the dam' forces, and other officials—stood opposite the location of the present clubhouse.

Stilson's Pond

Downhill, behind the present clubhouse was Stilson's Pond, at one time the reservoir for Gatun. It was named for Joseph H. Stilson, a "down Easter" from Maine; his father, Charles, had come to Panama in 1863 to work for the Panama Railroad.

During the dry seasons, he and his family lived in a big house, built by the French Canal Company near the old village, about where the center chambers of the locks are now. Miss Louise Stilson of Colon and two of her brothers, William and Joseph H. Stilson, Jr., until recently ticket agent for the Panama Line, were born in Gatun. Mr. Stilson, Sr., was in the hardware and lumber business in Colon.

When work began on the dam, the Stilsons moved to another large house, later destroyed by fire, on a high point of land near the present railroad. Stilson's Pond, on old pasture land, came into being when Gatun Lake was formed. When the Third Locks excavation was going on about 1940, the pond was filled with its spoil.

A few of old-time Gatun's street names have survived all of the town's changes. There are still Lighthouse and Schoolhouse Roads, for instance. Telephone Road, also known as Skunk Hollow, is now San Lorenzo Street; Front Street is Bolivar Highway; Santa Rita Place was once known as Hogan's Alley or Incubator Row.

Fun and Games

Life was simple in early Gatun, but people had fun. There was a woman's club, with Mrs. Chester Harding as its 1908 president. The men could belong to such organizations as the Inca Tribe of the Improved Order of Red Men which gave a ball and banquet on Thanksgiving Eve, 1907, with children in Indian costume attending. The Gatun Dancing Club met regularly, and occasionally, a touring company like the Edith Harris Scott Company gave performances at the clubhouse.

Men with outdoor bents, like Charles Thomas, played baseball on a diamond between the end of the lock wall and the

E. L. ROADES, Gatun Commissary Manager

present station, in an area long since under water. Or they could hunt tigers and red mountain lions across the Chagres as Charles H. Bath, now of Margarita, did frequently. On hot Sundays it was possible to borrow an engine and railroad car to ride to the beach at Chagres.

By March 1913, the population of Gatun was 8,887. Nine months later it had dropped to 5,943. The dam and spillway were finished, the locks were operating, and only clean-up work remained. An official estimate of that time gave the future population of Gatun as 160 American employees and their families.

Housing Replaced

Except for the introduction of Army troops into Gatun during World War I and some talk a few years later of abandoning the whole town, nothing much happened to Gatun until 1928, when new quarters were built for 164 local-rate families. In 1932, plans to replace most of Gatun's old housing were approved and grading for the $1,250,000 project began January 31, 1934. Buildings came down right and left. Even the old police station was demolished. Its officers set up temporary headquarters in a small frame cottage but transferred their prisoners to the sturdier jail at Cristobal.

Hardly had the new town been finished when Gatun went through another of its recurrent upheavals. The Third locks project which had been under consideration to some extent since about 1930 finally became a reality. On August 11, 1939, Congress authorized the immediate construction of the third locks.

Island Between Locks

At Gatun this meant the building of a new triple flight, each chamber 1,200 feet long and 135 feet wide. They were to be located about half a mile east of the original

LT. DAVID B. MARSHALL, Fire Station Commander

flight. Gatun was to become an island between the two sets of locks and was in for some of the greatest boom days of its up-and-down history. An official estimate of the force to be required set a peak of over 9,000 workers by 1943.

In January 1941, the contracting firm of Wunderlich & Okes signed a contract for the Gatun excavation. Construction men moved in. In the bottom of the third locks cut, now a great, gaping hole, giant shovels dumped their loads into dozens of trucks which raced about on the right-hand side of the imaginary highways below, and then, when they reached the top, switched over to the left-hand drive and sedate speed limits of those days. From an observation platform, which still stands at the end of High Street, anyone could watch the ordered turmoil below.

A few months after Pearl Harbor, Samuel Rosoff of New York, won the $45,705,000 contract to build the new Gatun Locks. Wunderlich & Okes completed their contract in May 1943, but the Rosoff contract was canceled. Shipping had been diverted to the war areas, cement and steel were all but unobtainable and there was military difference of opinion on the strategic value of the third locks.

War Days

With the war, the physical appearance of Gatun changed. Solid 26-foot fences of corrugated metal surrounded the lock area. Barrage balloons were anchored overhead. Buildings or part of buildings which might be fire hazards and, burning, light the vital locks target, were torn down. Air raid shelters were built and air raid drills held. Like all other Canal towns, lights were out by 11 p. m., there were no street lights, and cars drove with blacked-out headlights.

As the war receded into the Pacific and danger to the Isthmus abated, Gatun—and the rest of the Canal Zone—went back to its normal way of life.

On March 31, 1944, just 35 years after its first clubhouse was built, Gatun's present clubhouse was inaugurated formally. It was called the "newest and most complete of any in the clubhouse system." About 40 Zonians who had lived in Gatun in 1910 were invited to the dedication. Some of them--Lawrence Adler, Roy Dwelle, Reed E. Hopkins, Sr., and Charles E. Thomas - are still on the Isthmus.

Gatun Today

Today Gatun is a town of about 2,160 people. Its U. S.-rate commissary and clubhouse and post office are under one roof. The local-rate commissary and clubhouse are also combined, physically. There are two churches in Gatun proper, several in the local-rate section of town which is generally known as Chagres.

Gatun has an active Little Theater group and its residents think that it has more hobbyists than any area of like size.

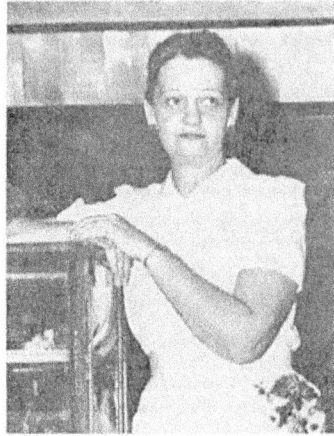

MRS. EVA REED, Gatun Clubhouse Manager

MARTIN S. SAWYER, Postmaster

CLIFFORD GREEN, Chagres Commissary Manager

OSMOND N. DUVERNEY, Chagres Clubhouse Manager

The grind of the power saw is a familiar sound. Camera enthusiasts, shell, coin, and stamp collectors, dog fanciers, and icthyologists abound. Several well-known local artists have lived there or still do. One Gatun woman is the author of a book of children's fairy tales.

Fishermen come from near and far to its Tarpon Club, beside the Gatun spillway where there is some of the best fishing in the world. Its town barbecues are famous. They are good, old-fashioned affairs where the men dig a deep pit, and work all night turning a beef or a pig over red-hot coals to the proper degree of rich, brown crispness.

There are often community picnics or dinners at the Block House, another Gatun institution, and its active Civic Council always arranges festivities for Christmas, Hallowe'en, and Fourth of July. The Christmas decorations which are an annual feature on the locks have inspired the townspeople to similar, if less elaborate, efforts and a drive through Gatun during the holiday season is well worthwhile.

Carl Nix, who works at the Gatun Hydroelectric plant, is president of the Gatun Civic Council. Although he is a transplanted Pacific sider, he now considers Gatun the best place in the Canal Zone.

"It's the friendliest town on the Isthmus," he says. "It doesn't make any difference whether you've been there ten days or ten years. You're part of Gatun."

Interior of the Commissary, Circa 1910. *From the Hallen Collection.*

HORSES

The Panama Canal Review, Fall 1975, pp. 23-28.

In the Canal Zone horse world there are hundreds of horse owners and riders. There are ten Riding Clubs conveniently located throughout the area.. It is all family fun and friendly competition for the horse owners and riders at the local horse shows. The Gatun Saddle Club is one of these ten clubs. "No hooves, no horse" is a well-known saying among the horse crowd. The riding clubs' goals are to have healthy horses, good horsemanship, and companionship.

The presentation of the colors at the Gatun Horse Show on the banks of the Canal. *The Panama Canal Review*, Fall 1975, page 27.

Waters of Gatun Lake cover Isthmian townsites

The Panama Canal Spillway, July 3, 1985, p. 3.

While there are no known mysterious civilizations lying beneath the surface of the Panama Canal such as the fabled island of Atlantis in the Atlantic Ocean, there are many towns and villages of the construction era covered by the waters of Gatun Lake.

Matachín, at the south end of the lake, was the home of 1,000 Chinese laborers in the early 1850s. They were brought to Panama to help build the Panama Railroad. Melancholia and the inability to adjust to the environment led to mass suicide and a high death rate from illness. Within a few months only 200 were still alive. The survivors were sent to Jamaica and prospered as farmers and businessmen.

Matachín appears on a map drawn in 1678 by Esquemeling and published in his book "Buccaneers of America." Also included on the map are the towns of Caimito, Cruces, Barbacoas, and Bailamonos, all of which are now beneath Gatun Lake.

Ahorca Lagarto, another construction day town, is also beneath Gatun Lake. Its name has its origins in a 16th century battle. Cimarrones, a group comprised of escaped African slaves and Panamanian Indians, were terrorizing travelers along the Camino Real. A regiment specially trained in jungle tactics were sent to deal with the outlaws. Their ensign was a lizard, which was worn on each soldier's helmet and breastplate. The Cimarrones, seeing the ensign, took it as their battle cry, "Ahorca Lagarto," which means "Hang the Lizard." The regiment was wiped out and the Indians in the nearby village used the battle cry to refer to the spot.

The place retained the name for 300 years. It became a station on the Panama Railroad.

Bohío Soldado, which means "Soldier's home," was the proposed site of one set of the French Canal Locks. John L. Stephens, the original owner of the Panama Railroad, had a cabin between Bohío Soldado and Frijoles. The cabin site is at the bottom of Gatun Lake.

Lion Hill was named by Panama Railroad construction workers who said that the din sent out by the monkeys there sounded like the roar of lions.

Residents in the lake villages were averse, for the most part, to moving out of their homes. One resident ventured his opinion that the waters would not come, since "the Lord promised never again to flood the earth." Others remembered that such impossible things had been spoken of by the French 25 years before and had never happened, so they saw no reason for leaving. But the rising lake level proved them wrong and led to the relocation of the villages' inhabitants, their possessions, and in some cases, their houses.

What's Happening

The Panama Canal Spillway, December 18, 1987, p. 2.

On Christmas Eve, Santa makes his annual stop at the Gatun fire station around 7 pm. Santa arrives aboard his truck-drawn, red and white sleigh. Parents send a gift for their child to the fire station, wrapped in plain paper with the child's name written in bold letters. On Christmas Eve, children are excited to see Santa and receive a gift from Santa.

The Tarpon Club, located on the west bank of the Canal near the Gatun Hydroelectric Plant, hosts a New Year's Eve party, beginning at 9 p.m., December 31. Included in the price of the ticket will be live music, a buffet, open bar, champagne and breakfast. Dress is open to anything from jeans to tuxedos, and everyone is invited.

The Gatun Community Youth Center, sponsored by the Employee Fitness Branch, has planned several events for this month. Christmas caroling begins at 7 p.m. on Monday, December 21. Plans for Tuesday, December 29, include sand sculpting and an all-day beach party. And, if there's enough interest, on

Thursday, December 31, there will be a New Year's Eve party for teens.

The restoration committee of the Gatun Church of the Immaculate Conception, Catholic Church, is pleased to announce that weekly Masses will be celebrated at 5 p.m. on Saturdays.

The restoration project of the 50-year old church is completed, and everyone is invited to the services. A Christmas Eve Mass will be celebrated at 11:45 p.m. on Thursday, December 24. The church is located on Bolivar Highway in Gatun.

Memories

Red Feather Hill, Gatun

At the entrance to Gatun there was a hill that stood out above New Street, and looked down on the Elementary School. We called this hill Red Feather. It was a magical place where we could see for miles to the Gatun Locks, to Ft. Davis and all the way to Colon. A gravel road led up Red Feather. It was used by residents for star gazing, a lovers lane by teens and by the army as a tank location during maneuvers. Red Feather was a place from which the army could conceivably protect the Canal. A manned tank appeared there numerous times a year, and as kids we always paid them a visit. We bothered them to share their K rations, show us their weapons, and climb on their tank, all requests that were granted often enough that we always asked!

The other attraction of Red Feather was the steep hill in the front of it facing New Street that was bare of grass and its bright red dirt scarred with erosion that left gullies in the hillside creating feather shapes, hence possibly the origin of its name. This short slope was the source of much consternation to our mothers as they tried to remove the red clay dirt from the bottoms of our pants because, you see, every visit to Red Feather involved sliding down these feather shaped slopes!
- *Antonia "Toni" Klasovsky Treverton*

Frijoles: Boat Rides and Rubber Balls

In 1940 I lived in a little known settlement on an isolated peninsula of Gatun Lake where my father was sole agent at the tiny railroad station. Accessible only by boat or rail, it served as a transfer point for the shuttle boat of the Smithsonian Institution's Tropical Research Facility on Barro Colorado Island. There were only two houses in that "town."

When not commuting by train to school in Balboa, I was often on Gatun Lake in a trader's boat as it motored to the native villages scattered around the lake. I observed with fascination, villager's activities, most notable, the curing of raw rubber. A villager would dip the tip of a stick into a container of latex harvested from local trees and turn the stick over a smoky open fire. The dripping stream of latex became solid as it changed from white liquid to a soft brown thread of rubber creating a heavy, tightly wound ball, the size of a softball. Bundles of these were sold to traders and one given to me. It had a fantastic bounce and even floated.

Years later, I demonstrated this "super ball" quality at Cristobal High School. I hurled the native rubber ball from the second floor balcony over front entrance, and it bounced almost all the way back up to me. The attending disturbance resulted in the confiscation of my prized possession, to be locked forever in a cabinet in the CHS science room.
- *Fred Raines*

Benevolent Canal Zone Police

One evening while home on leave from the Air Force I decided to take a girl I liked out for a drive. Looking for a spot where we wouldn't be disturbed, (All our towns had one or more "lovers' lanes.") I drove down a jungle trail near the Gatun tank farm. The road was narrow, really just tire tracks in the bush, and I backed off the main trail about 50 feet further just to be sure we would have some privacy. While "taking care of business," I had a funny feeling and looked over the dash to see lights coming down the main trail. My heart jumped when I saw a Canal Zone police car drive past my little private necking spot. Figuring the policeman had seen my car but also knowing he had to drive to the end of the main road before turning around, I decided to make a run for it. Speeding out of the jungle without lights, I tried to lose the policeman in the streets of Margarita. No such luck. The red light came on, and I was busted. The policeman came up to my car window and looking in said, "Gerry?" Looking back I said, "Andy?" It was my cousin. "What were you doing back there in the bush?" "Nothing," I said sheepishly. Putting his ticket book back in his pocket, he simply said, "Take that young lady home and get home yourself." And with that we both went our separate ways.
- *Gerry DeTore*

Gatun Side Yard: Baseball and Bundles

As the whistle of the afternoon train signaled its approach, my friends and I cleared baseball gear from the platform adjacent to the railroad tracks of the Gatun railroad station where we had been playing. We called that platform my "side yard" because the upper level of the station was my family's living quarters. The oncoming train would bring bundles of the evening newspaper, The Panama American, which we pre-teen "newspaper boys" would deliver to Gatun homes.

As the train came to a stop at the station, my Dad, the Station Agent, was in animated discussion with the train's Conductor, gesturing toward the second class passenger coaches and fuming that the train was delayed in its departure. As we opened our newspaper bundles, I heard my Dad on the phone with the Dispatcher explaining that a woman on the train had gone into labor causing a delay.

Many years later, a newspaper reported that the new Panama Canal Railway Company had named a locomotive after the great former baseball player, Rod Carew. The article also described the unusual circumstances of his birth on October 1, 1945. Only then did I realize that when a first class passenger, Dr Rodney Cline, was summoned to the second class coach of the train stopped at the Gatun railroad station to assist in the birth of a baby boy, newspapers were not the only bundles delivered that day.

Rodney Cline Carew, Hall of Fame baseball player, was born on a train in my Gatun "side yard."
- *Fred Raines*

Gatun Elementary School.
Courtesy of Andy Nash English.

Above and Below, Costumed 5th and 6th grade students at Gatun Elementary in 1945. *Courtesy of Fred Raines.*

Gatun Union Church.
Courtesy of Andy Nash English.

Gatun Union Church Junior Choir, 1946.

Lighthouse
Courtesy of Andy Nash English.

4th of July parade. *Courtesy of Elaine Stevenson.*

Gatun Pool and Gym. *Courtesy of Andy Nash English.*

Gatun Clubhouse, circa 1910. *Courtesy of Robert Karrer.*

Right, A Halloween costume party at the old Gatun Clubhouse in the early 1920's. Ruth Egolf Clement is in the third row at the left in a tiered dress.

Below, The old Gatun Commissary in the background and Horine-Obold-Egolf family members in the foreground in 1918. On the left are shown Emily and Paul Carlton Horine with Emily's sister, Mary Obold, who were visiting the Horines' daughter and family, living in Gatun and shown on the right--Leon and Ruth Egolf with their two children, Harry and Ruth. Mary liked the Canal Zone so much she also decided to move there to work and live. A brother of Ruth's, George Carlton Horine, also came from Reading, Pennsylvania, with his wife, Esther, to live and work in the Canal Zone. The Egolf and Horine families each had five children raised in the Canal Zone; all five of the Egolf children stayed in the Canal Zone as adults and worked for the Panama Canal Company. *Courtesy of Barbara Egolf Dedeaux.*

Gatun children with Joe Ebdon, Sr., about 1920, perhaps in front of the old Gatun Clubhouse. The three children at the right are unidentified. The others, left to right, are Ruth Egolf Clement, Doris Ebdon, Fred Ebdon, Harry Egolf, and Joe Ebdon, Jr. *Courtesy of Barbara Egolf Dedeaux.*

Gatun, CZ, December 7, 1941

It was December 7, 1941, and I was living on the Atlantic side in the town of Gatun. Our house was the closest house to the Gatun Locks and it was a duplex on stilts. It was located near the train station that connected the Atlantic and Pacific sides of the Panama Canal. Gatun was the Gatun locks town where most of the Atlantic side U.S. Locks employees resided. On that day, the U.S. entered World War II when Japan bombed Pearl Harbor, and the residents were out in the street, looking at the sky, responding to rumors that there might be sabotage or the Spillway might be bombed. The Spillway if bombed would mean the water would drain the Gatun Lake, thereby making the locks non-functional. My mother was worried because my father was out on the lake and had not returned home when the news broke. At that time, he was a policeman and had to report to duty. When he returned he immediately went to the police station and began his daily shift.

The military was on alert around our house and were "trigger happy" and suspicious of people around the locks. I attended Gatun Elementary School, and one sentry yelled at me, "Halt, who goes there?" and I was only in the second grade. During the next week or two there were air raid alerts; my father had to go to the Canal Zone Police Station at odd hours, and at that time the police were not paid overtime.

The US Marines set up a tent city right outside the Gatun Elementary School. Each morning when I went to school, I used to walk by the tents and the marines who let us play with their equipment on the way to school and home from school as well.

My father took me to the Gatun Police Station so I could see the Italian Officers off an Italian ship that was confiscated as it arrived at Cristobal. The officers of the ship were interned as prisoners of war for the duration of the war in the Gatun Police Station.

Another time my father took me to see the survivors of a torpedoed ship as they were coming ashore covered with oil in Cristobal, the dock area of Cristobal Pier. These were memorable and historic times in Gatun.
- Jerry Halsall

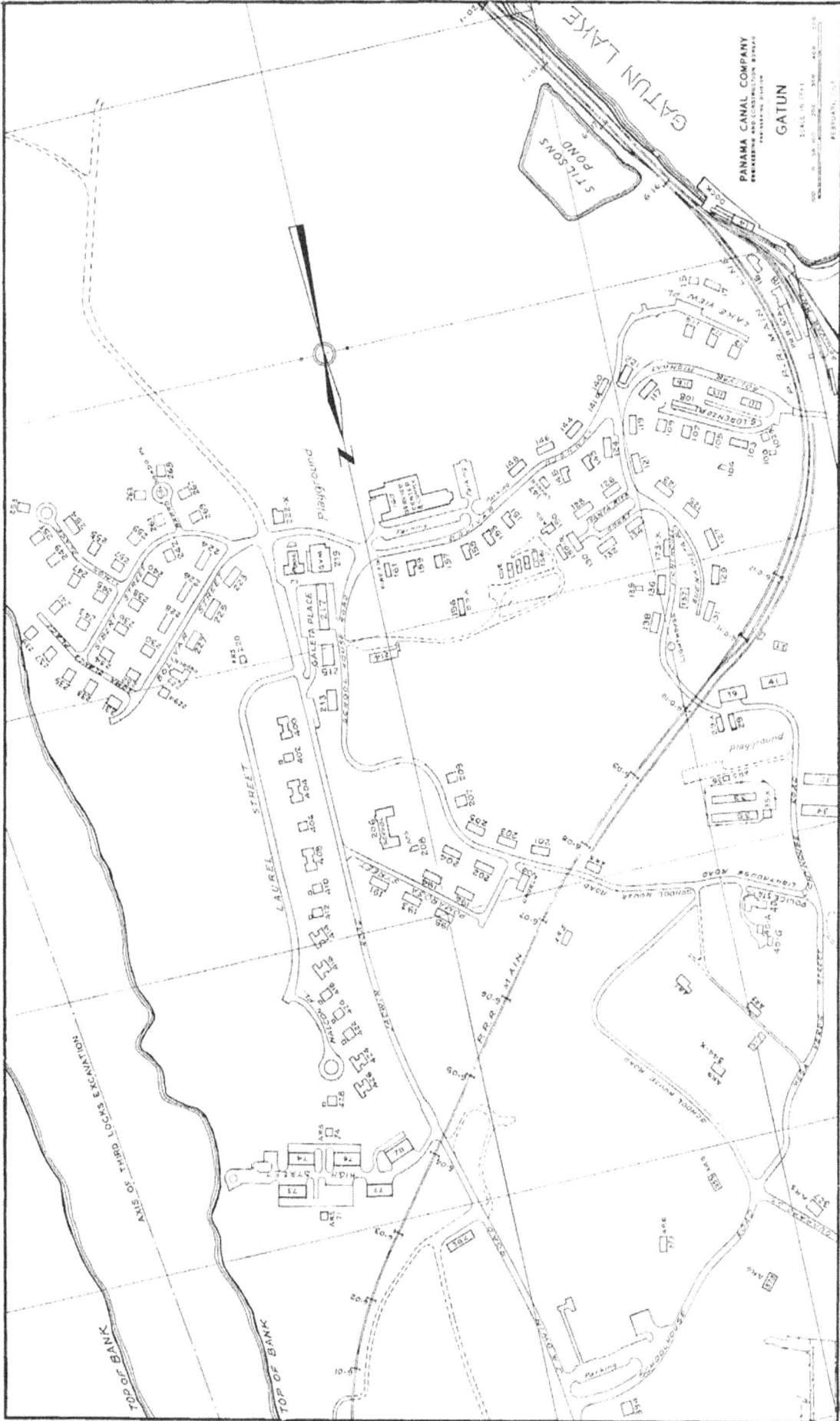

PANAMA CANAL COMPANY
ENGINEERING AND CONSTRUCTION BUREAU
ENGINEERING DIVISION

GATUN

GATUN LAKE

STILSONS POND

Photos Show Progress of Quarters Program

The Panama Canal Review, June 6,1952.

The quarters construction program this year rivals that of any year since the close of the Canal construction period. Not all of the construction sites are pictured here.

The first photo shows construction of 9 new apartments in Rainbow City. A total of 131 masonry houses are to be built at Margarita (second photo) in the northern extension to the present town. Twelve new masonry houses are going up in Diablo Heights (center photo) on a bluff overlooking Albrook Field. In Gatun, Isthmian Constructors are building 10 houses on Jadwin Road (fourth photo). The largest of the Pacific-side projects in the US-rate communities, is that at Ancon (fifth photo) where forty-eight new quarters, to house 71 families, are rapidly taking form.

The 1950s were excitng times as new housing began to spring up throughout the the Canal Zone, giving families something to dream on.

Second
Housing
Construction
Period
1950s

La Boca

Sponsors

Tamara Martinez Gramlich
Peggy (Hale) Huff

Your Town - La Boca

The Panama Canal Review, June 4, 1954, pp. 8-9

If a section of Panama Railroad track had not sunk six feet one morning in 1907 the Canal Zone town of La Boca—"The Mouth"—might not be where it is today. The La Boca area might look like the environs of the two-step locks at Miraflores instead of what it is—one of the Canal Zone's oldest local-rate towns!

The canal plan had called for two sets of locks, one at Pedro Miguel and the other near Sosa Hill. They were to have been separated by a large terminal lake, to be known as Sosa Lake. Not all the Canal's top men—John F. Stevens, among them—approved the idea but had begun work on the dams for the lake.

After the section of track near La Boca sank suddenly and a trestle toppled, Chief Engineer George W. Goethals appointed a board to study lock sites. Eventually the present locations were determined and La Boca returned to its former status of Pacific terminal for the Panama Rail-road and the only Pacific port between Callao in Peru and Salina Cruz in Mexico where deep-draught vessels could unload at a wharf. It seems strange today to read that the transfer provided not only more stable foundations but also better protection from bombardment from the sea!

French Days

As far as its history goes, La Boca went through three phases. At La Boca the old trail from Panama City to the towns which are now considered to be in the "Interior" crossed the Rio Grande. The French Canal Company, as the Americans did later, used the valley of the Rio Grande as the southern end of their canal line. In 1881 they began to build shops at La Boca where their dredges could be assembled.

One historian reports that the French Company loaned enough money to the Panama Railroad for construction of a deep-water harbor and a 960-foot steel pier. This pier, which eliminated the old lighter system, is still standing although it has been much changed in appearance. Just as the Americans did later, the French planned for a double lock near La Boca.

When the French Company sold its interests to the United States in 1904 the buildings and wharves in La Boca were part of the properties transferred. No better

LA BOCA, first Canal town seen from ships entering the Pacific end of the Canal, is at almost the southernmost tip of the Canal Zone. A small bit of the ocean is visible in the upper right. The school buildings, where over 1,000 pupils attended classes last year, are at the top of the photograph.

description of this phase of La Boca can be found than the following, from the 1905 report of the Isthmian Canal Commission:

American Days

"The town is divided into two parts by the railroad tracks and yards. On one side all of the buildings are owned by the United States and on the other nearly all of the buildings were erected by private parties on land leased from the old French Company. All of the buildings in this town owned by the United States are being overhauled and repaired; several of the more dilapidated were destroyed and in their places have been erected two large and commodious barracks, one for the unmarried and one for the married employees working at this point.

"Repairs on the old ones have reached such a point that it is proper to say that this portion of the town has been rebuilt and instead of being a dangerous plague-spot, the town has now become a model camp with houses in good repair, freshly painted, supplied with electric lights, a water system and good drainage. A good road of Telford pavement constructed by the Commission connects La Boca with the outskirts of Panama."

This was written after two cases of bubonic plague had broken out at La Boca. The resultant quarantine disrupted the transportation system and called for stringent measures by sanitary authorities.

At this time La Boca was primarily a settlement for American employees. Its commissary, opened in September 1907, supplied Ancon and Balboa by wagon each morning; current from its electric light plant was furnished to Ancon and Balboa and, later, to Corozal.

There was an elementary school, located about where the present Service Center stands. In September 1908, however, THE CANAL RECORD reported: "Owing to the small number of children attending school at La Boca last year, that school has been abolished and the children at La Boca will be transported to and from the school at Ancon in a wagonette."

This phase of La Boca history ended in April 1909, when, by executive order, the town was renamed Balboa. The Peruvian

Minister to Panama had suggested the change, saying "As the Atlantic entrance to the Canal is named 'Cristobal Colon' for the great navigator and discoverer of our continent, so should the Pacific entrance be named after the intrepid Balboa, its discoverer." Thereafter, although there continued to be a thriving Pacific side town it was known as Balboa or East Balboa, and the name La Boca disappeared from official records, temporarily.

Again La Boca

In August 1913, exactly one year before the S. S. Ancon made the first official transit of the Canal, Colonel Goethals authorized "the construction of a labor camp at La Boca to provide accommodations for West Indian laborers." At this time the name for the town had not been chosen.

Several names were suggested: La Boca, Lesseps or deLesseps; Espinosa, after the founder of Old Panama; Morgan Town, for the buccaneer; and Lincoln, in honor of the Civil War President. Some objection was found to all except the first of these and on August 18, 1913, the town was officially named La Boca.

La Boca, substantially as it is today, was laid out in a rectangular plan on part of a large fill, southeast of Sosa Hill. Both streets

R. D. MELANSON
Commissary Manager

MRS. BEULAH M. SHOEMAKER
Nurse-in-Charge, First Aid Station

and avenues, which now bear such names as Martinique and Granada Streets and Jamaica Prado, were originally numbered. The town was divided lengthwise by a park: family quarters were all on the south side and bachelor quarters on the north. A commissary, near the present entrance to Dock 6, supplied "canned vegetables and cold storage goods."

An elementary school was opened early in 1914 and in September of that year its enrollment was 129. Last year over 1,000 students attended the three La Boca schools. Street car tracks ran parallel to La Boca Road and provided transportation to Panama City. The street car service was maintained until 1941.

Houses – Old And New

Some of the houses were new but many were brought from towns 'which were being abolished. There were the old hospital, dispensary, and commissary from Portobelo, converted into living quarters; a laborer's barracks from Gorgona; a barracks from Paraiso; two houses from Gatun; and several from Diablo. Not all of them had plumbing or kitchens. Washhouses, outhouses, and cook sheds stood behind each small group of quarters.

Wood to fuel the stoves was dumped daily into boxes at the entrance to each set of quarters by a wagon from a sawmill located where the Balboa police station now stands.

Maintenance of La Boca was something of a problem. In late 1914, the administration offered prizes of $5, or a percentage reduction in rent, for the best-kept, neatest, and cleanest quarters of various types. The prize system was followed for about two years.

While La Boca was primarily a town for local-rate workers, there were a few American families living there. Most of the Americans were people waiting completion of quarters in Balboa and Balboa Heights. Old La Bocans still call the street where they lived "Gold Street."

In 1915 the Acting Governor, Chester Harding, turned down a Metal Trades Council request that the La Boca quarters be assigned to Americans and said: "The administration hopes to provide quarters sufficient to house every Gold employee on the permanent force. Quarters are being constructed as fast as the money is available. I think the present unsatisfactory state will not continue for more than a year."

Sports Town

Present day Zonians know La Boca as a great sports town and it was in its hey-day in the 20's and 30's. The East End cricket team, later the La Boca Cricket Club, drew big crowds to their Sunday games and mid-week games were scheduled whenever a British ship came into port.

Both men and women played "rounders," a sort of ballgame, between their quarters, and the youngsters diverted themselves sliding down Sosa Hill—nowhere near the same shape now as it was then—on palm fronds, or swimming in the sea behind the old Clubhouse. Dominoes, which can be quite an athletic sport when played in La Boca, was an under-the-house pastime.

Goats, kept for their milk, wandered through the town and bees from the

FRANCIS A. CASTLES
Principal, La Boca High School

numerous apiaries—some families had as many as 30 hives behind their quarters—stung unwary passersby. Fishermen, in those "good old days," sold their catches not by the pound but by the string and for as little as 10 to 15 cents a string. Movies cost 5 cents for youngsters in the silent film days.

EDWARD A. GASKIN
Principal, La Boca Elementary School

The war boom almost doubled La Boca's population. From 3,228 in 1930, the town swelled to 6,076 in 1942. Contract laborers from Colombia, Salvador, Costa Rica, and the West Indies were housed in giant new barracks and fed from central messhalls.

As nearly as possible they were separated by nationalities-Jamaicans in

CECIL C. GITTENS
Service Center Manager

one building, Colombians in another, etc., but despite this there were international difficulties which flared into frequent disturbances until the men learned to work together and keep their frazzled tempers under control. Sports helped to make them friends—although international rivalry was keen—and the various groups formed football teams which drew crowds of two to three thousand at their matches.

Facilities Added

During the hectic World War II period the town's facilities were strained. A theater was built as an addition to the Clubhouse; the restaurant was enlarged. A library was established in the high school which had been built in 1937 and a dispensary and emergency fire station were built. Town spirit made itself felt in an active Civil Defense group. After the end of the war the contract laborers went back home and the town returned to normal, except for such flurries of excitement as the Commissary fire in February 1950. An emergency commissary was established in the old restaurant but it was months before the badly-damaged Commissary building was repaired and back in operation.

1953 Population About 3,000

A year ago La Boca was a community of about 3,000 and its population is still dwindling as its people are moving into newer and more modern quarters in Paraiso. But there are still six churches in the town.

It still has its big Commissary and one of the largest Service Centers—La Bocans still call it a clubhouse—in the Canal Zone. Headquarters for the Balboa Federal Credit Union are in the Service Center building and its auditorium, once the movie hall, is used for public meetings. The building also houses a barber shop, beauty shop, and shoe shop and a small private typing school is operated in its basement.

The Lat-teen Club, a junior organization, is back in the old restaurant building from which it was hastily evicted when its space was needed for the emergency Commissary. This club provides a recreation spot for the senior juniors of the town.

Dominant features of La Boca are the two school buildings. During recess periods, between classes and just before and after school, that end of town is a hubbub of activity.

Old La Bocans are proud of the fact that their town has not lost its reputation as a "great town for sports." Cricket is back again and one fan holds that this year's league is the best yet. There is still football—of the soccer variety—baseball and softball. Outstanding athletes, like Carlota Gooding who recently won the 100-meter race in the Central American Olympics, and Frank Prince, who won the 800 and 1,500 meter events at the same Olympics, are products of La Boca training.

No story of La Boca, the townspeople say, would be complete without mention of them and the man who trained them and other youngsters—Ashton Parchment, La Boca physical education director.

La Boca Once Largest C. Z. Town Will Soon Become Just a Memory

The Panama Canal Review, September 6, 1957, p. 4.

La Boca, one of the oldest and for many years one of the most populous towns of the Canal Zone, is to join a long list of former towns at the end of this year.

The community is now down to two lonely rows of frame buildings, a population of less than a fifth its former size, a scant half dozen public buildings, and a big playing field where 40 years ago cricket matches lasted out an entire Sunday.

Plans for the abandonment of the town have been stretched out now for several years since the extensive quarters construction in Paraiso.

Nearly half of the La Boca families will be assigned quarters at Pedro Miguel in the 12-family apartment buildings which have been recently occupied by Air Force personnel. Eighty-four families can be accommodated there and about 30 others at Paraiso or Santa Cruz. About 50 families will move to Panama, and most of the 125 bachelors will be required to seek quarters outside the Zone.

As soon as all the remaining quarters are vacated, they will be demolished. Some of the oldest buildings in use were erected in 1910 and were rebuilt in 1914. Most of the others were built in 1930.

At this time no definite plans have been made for the permanent use of the area or of the public buildings which will remain. La Boca Commissary, housed in the town's only masonry building, and the Service Center, once the largest of any Latin-American town of the Canal Zone, will be closed at the end of the year.

New Town Rising on Old Site

The Panama Canal Review, March 6, 1959, p. 13.

The revival of La Boca as a Canal Zone community is starting this year with the installation of a new street and municipal facilities, and construction of five masonry houses.

The site of the new construction is on La Boca Road just east of the former Commissary.

The new community is scheduled for completion during 1961 when it will have 33 one-family and 30 duplex houses.

The homes being erected in La Boca are replacements for family units in Balboa which are to be demolished for the construction of the bridge.

During the coming fiscal year, 45 family apartments will be built with another 41 apartments to be built in the following year.

The construction at La Boca is part of The Master Housing Plan, which calls for the elimination of all substandard housing for the permanent U.S. citizen employees of the Canal organization.

Some of the new houses to be built in La Boca will follow this design.

Type 350 was a 1,254 sq. ft. masonry single family with three bedrooms and two baths. Similar homes were also built in Ancon and Balboa. *Courtesy of Ed Ohman.*

Showing new Building at La Boca.

This vintage postcard shows the style of
housing being build in La Boca during the
construction era.
Courtesy of Robert Karrer.

Below Left, Canal Zone Post Office,
circa 1904. Below Right, Police Station,
circa 1904.
From the Hallen Collection.

Canal Zone College Opens

Schooling in the Canal Zone, 1904-1979.

The new campus of Canal Zone College opened
on September 6, 1963. It had just received accreditation for a three year program and opened its doors to
not only Canal Zone residents but extended enrollment
to Panamanians as well. Enrollment tripled causing
immediate overcrowding and a shortage of textbooks.
"The textbook shortage forced College officials to conduct a campaign among the alumni to contribute or re-
sell used textbooks through the Panama Canal Supply
Division, It was necessary to make special airmail orders from the United States at a considerable additional expense. It wasthis situation that spawned the idea
of a bookstore operated by the Studend Association,
which has proved successful up to the present time."

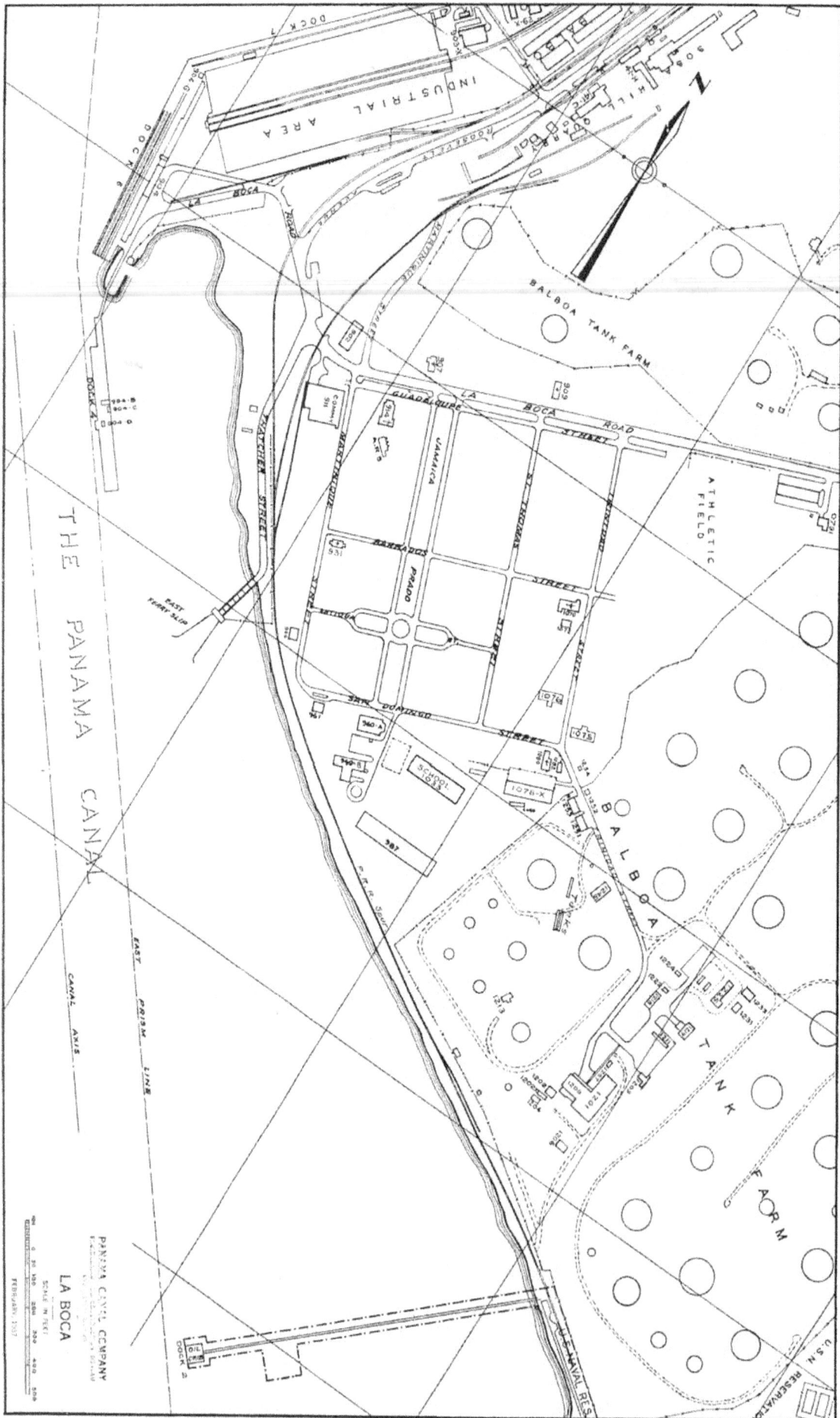

PANAMA CANAL COMPANY

LA BOCA

SCALE IN FEET

FEBRUARY 1957

Los Rios/Corozal

Sponsors

Lester, "Chefa" & Lena Barrows
Dennis & Peggy Huff
Bob & Cheryl Russell
The Lynn Stratford Family
The Wards—Marvin, Jackie, Debbie, & Lissa
Joseph H. White, Jr., & Family

Your Town - Los Rios

The Panama Canal Review, December 2, 1955, pp. 8-9

LOS RIOS, newest of the Canal Zone's towns—received its first residents two years ago. The town layout, with its surrounding street, loops, and deadends, is new in Zone planning.

Los Rios, newest of the Canal Zone's communities under its present name, is just what its predecessor, Corozal, was half a century ago-a 'commuters' town.

During the early years of the twentieth century. Corozal was a cluster of houses and a railroad station. Los Rios is a cluster of houses-albeit a larger cluster than in 1904-and a railroad station. It has no commissary, service center, post office, or school. There is not a shop, nor office activity in the whole town. And that was just exactly the situation in 1904 when the United States was beginning to build the Panama Canal.

The Corozal of 1904, like the Los Rios of 1955, was connected to Panama City by railroad and by highway. But the road of 1904-it could hardly be called a highway-was generally impassable except for pedestrians or saddle or pack animals.

So it was by railroad rather than road that Corozal's first American residents got back and forth to their jobs, most of which were in the Isthmian Canal Commission's headquarters building in Panama City.

Boxcar Quarters

Housing in the city proper was almost nonexistent; the few buildings in Ancon belonged to the hospital; Balboa of those days was La Boca and its few structures were overcrowded. Boxcars were hauled to Corozal and parked on sidings and they and the village's two dozen or so French buildings were reconditioned for dwellings. One of Corozal's first commuters was C. A. McIlvaine, who later, and for . many years, was the Canal's Executive Secretary.

These few buildings were nowhere nearly sufficient for the rapidly growing force, so in 1905 the Commission built at Corozal a mammoth, three-story hotel which provided room and board for its more than 100 bachelor occupants. It is difficult to orient the hotel in today's surroundings as there was no road in those days beyond Corozal toward Miraflores. It stood, some old-timers say, somewhat north of the present railway station and on the east side of today's highway.

William Weigle, who served with the Sanitary Department for many years and whose son now lives in Margarita, used to tell a ghoulish story of his assignment to the Corozal hotel. At first, hotel attendants said there was no room available. Finally one of them led him to a room on the second floor which still contained the belongings of its previous occupant, who had died of yellow fever a few hours earlier. Mr. Weigle did not spend that night in his new room. Instead he walked up and down the railroad tracks all night, waiting for the morning train to Panama City and other accommodations.

Dams And Locks

Had early plans for the Canal been followed, there would have been no Los Rios as we know it today and probably the history of Corozal would have been much briefer than it is. One of the earliest American schemes proposed a dam from Ancon to Corozal-one of a group of three- to back the waters of the Curundu River and other streams into a terminal lake above La Boca locks. But La Boca turned out to be unsatisfactory as a lock site and the whole project was moved about three miles inland to Miraflores.

One of the incidental moves which came along with this change of plan was the transfer of the Pacific Division headquarters from Sosa Hill to Corozal in December 1907. The long low building which, years later, served asthe Army's Finance office, stood on a knoll just about where the lower end of Los Rios is now located. The Division covered the entire southern end of the Canal work, from Pedro Miguel to the Pacific entrance; although a good many of its force lived in Corozal they fanned out during working hours to anywhere' from Miraflores to La Boca.

In 1908 Corozal consisted entirely of old French buildings, with the exception of the hotel; Division Engineer Sydney Williamson kept up a constant campaign for more and better quarters for his people. Eventually a few sets of quarters were built at . Corozal and other houses moved in from towns "along the line." In 1908 Corozal's population was 592, exactly 10 more than it was at the time of the latest Canal Zone census last fall. In 1908 the town became quite modern; electricity was installed in all quarters, messes, labor camps, and the hotel, and its people no longer could say that they were "lighted by fireflies."

History In The Making

More than one bit of Canal Zone history was made in Corozal. In 1905 it was the scene of the Zone's July Fourth celebration. As the Star & Herald put it, the Zonians "heard Judge Herman Gudger orate, witnessed a few races, and uncorked some fresh U. S. enthusiasm."

On January 2, 1906, the first free public school under the Canal Zone government was opened at Corozal. According to the Isthmian Canal Commission report, it was "provided only with a few borrowed chairs and tables and such books as had been sent to the school authorities as samples, supplemented by a few found in the homes at Corozal." Two years later a two-room school building was built at Corozal, but in 1910 Corozal children began to attend school in Ancon.

But probably the most important event for which Corozal was the setting occurred March 16, 1907, when the newly arrived chief engineer, Col. George H. Goethals made his now famous speech – "there will be no more militarism in 'the future than there has been in the past" –

FRENCH QUARTERS like these were the homes of the earliest residents of Corozal, now Los Rios. After the American canal builders arrived, these old quarters housed men working in Panama City.

before what historians reported as a "hostile audience." Years later, at a Society of the Chagres dinner at the Hotel Tivili, Goethals recalled that first speech at Corozal.

In 1910 a commissary alongside the railroad track north of the station was added to Corozal's small list of public buildings and that same year the YMCA which ran the big clubhouses in the major towns began operating the Corozal Club.

Music And Movies

Morris M. Seeley of Gamboa, who was right hand man to Dr. Dutrow at the Corozal dispensary during the construction period, recalls one clubhouse manager, George R. D. Cramer, whose eyes, one brown and one blue, fascinated the ladies of Corozal almost as much as did the manager's very fine voice. Mr. Seeley, who often played piano at dances and concerts, frequently accompanied him during the local talent shows at the clubhouse.

Mrs. Howard Fuller, who was a small girl in those days, recalls that she saw her first "talking movie" at that Corozal Clubhouse. The film was badly synchronized - with the recorded sound and frequently the actors were talking when they should have been listening.

Lock Forces

Corozal's peak as a Canal town – a population of a little over 2,000 – was reached between 1912 and 1914. As Gatun locks neared completion a large number of men working for McClintic-Marshall, contractors for the lock gates, were moved to the locks at Miraflores and were housed at Corozal. Several houses and labor barracks were converted to bachelor quarters for them and the old Culebra Lodge Hall, which had room for 112 men, was moved to Corozal.

Work was started at that time on a hospital farm which would care for the Canal's permanently disabled employees and its insane. The piggery, dairy, and poultry farm which had been part of Ancon hospital were moved to Corozal as part of the hospital farm, but these were discontinued some years later. Today, but with newer buildings and modern equipment, this section is known as Corozal Hospital. The Corozal cemetery was also established in 1914. From its original plot of two acres it has grown to its present 47-acre area.

Transferred To Army

The establishment of Panama Canal headquarters at Balboa Heights in 1914 and the construction of quarters in Ancon and Balboa put an end to Corozal as a Canal town. As of December 1, 1915, the village was transferred to the Army. Except for the replacement of the original wooden buildings at the hospital with the present concrete structures, the enlargement of the cemetery, and the construction of the quarantine station during the early years of World War II, there was no civilian activity in the Corozal area until about three years ago.

In March 1952, plans to construct a new townsite at Summit were abandoned and arrangements were made between the Canal organization and the Army to return to the Canal Zone government 100 acres of land in Corozal, east of Gaillard Highway.

Within a few months bulldozers were at work changing the face of the terrain and a few months later the first houses were going up. In October 1953, the first families moved into the new townsite. It was July 1954, however, before it was formally rechristened.

New Layout

The streets had already been named – all of them for Isthmian rivers. On the recommendation of the Pacific Civic Council, of which the town is a ward, the new town was named Los Rios, or The Rivers.

Los Rios' layout is something completely new in Canal Zone planning, without the checkerboard of crisscrossing streets of the other towns. It is surrounded by what engineers call a circumvallation road, or, in layman's language, one which follows the perimeter of an area.

From the perimeter three horseshoe-shaped loops and several dead-end streets extend toward the center of the area. There is not a major street crossing in the entire section.

Los Rios houses are the latest types to come off Canal engineers' drawing boards; a good many of them are the masonry, on-the-ground duplexes. A recreation area, complete with barbecue pit and bohio in addition to the more usual swings and slides and teeter-totters, is located in the center of the town-and is generally in use. Diablo Heights, which was a suburb of construction-day Corozal, serves the people of Los Rios at its commissary, service center, and post office and Los Rios children attend school at Diablo Heights or Balboa.

Model Railroad

No story about Corozal-Los Rios would be complete without mention of the model railroad which has occupied a room in the Corozal railroad station for some years. The railroad layout was built by the Canal Zone Society of Model Railroad Engineers but was sold last month, down to its last crosstie, to the Isthmian Model Railroad Association, formerly of Rodman. The new owners will leave their purchase in the railroad station but, for the time being, are not inviting visitors to see it.

Residents of Los Rios, like Mrs. E. G. Evans who is a town representative on the Civic Council, find a lot of advantages in living there. It is cool and quiet and there is no heavy traffic –

"Of course," she says, "we are always trying to improve it, but that's what councilmen are for, isn't it?"

LOS RIOS is represented in the Pacific Civic Council by this group. Left to right, seated, are: Truman H. Hoenke, Mrs. Evan G. Evans, E. W. Zelnick, and Charles A. Dubbs. Standing are Robert A. Stevens and A. M. Jenkins.

Corozal Fire Department and volunteers with hand operated rig, 1908.

LAYOUT IN COROZAL DEVELOPMENT REVEALS NEW CONCEPT IN TOWN PLANNING FOR ZONE

The Panama Canal Review, October 3, 1952, pp.1, 4.

A new concept in town planning for Canal Zone communities has been adopted for the development of the new residential area in Corozal.

In all, 128 single-family units and 20 two-family houses are planned for the area at Corozal which was transferred earlier this year to the Canal for a new housing project.

Instead of being developed by "blocks" with numerous street cross-ings in the residential area, one main boulevard will circle the site and short circle or dead-end streets will serve residences.

A large area will be kept free for an elementary school which is to be built at a later date. The only other community facility planned in the area is a swimming pool slated for later construction. The new houses in Corozal will be adapted to the terrain rather than hav-ing terrain graded to fit the houses. Where flat areas are not available for ground-level masonry quarters, elevated houses will be built.

Houses will be faced into the prevailing breeze wherever it is practical and all will be located to provide maximum space for outdoor living.

Isthmian Canal Commission Hotel, Corozal, circa 1910. *From the Hallen Collection.*

Hospital at Los Rios, circa 1919. *From the Hallen Collection.*

Corozal Cemetary. *Courtesy of Ed English.*

CHINESE GARDENS

The Panama Canal Review, January 6, 1956.

In 1906, the Isthmian Canal Commission received a letter from the Wah Me Hing Company of Baltimore, Washington and Hong Kong, asking permission to establish 10 or 15 truck gardeners in the Canal Zone. The first was to be located at Empire and the second between Mindi and Cristobal. By 1908, there were 42 Canal Zone land leases for Chinese gardens. (Photo shows Chinese Gardens on Gaillard Highway near entrance to Cardenas.) Gardeners operated on a share cropper basis; the bulk of their produce went to the commissaries but gardeners maintained one third for private sale. Most of the Chinese gardeners came from the Kwang Tung Province in south China, whose capital is Canon.

Chinese Market between Cardenas and Fort Clayton on Gaillard Highway. *Courtesy of Carl Berg.*

Memories

Living in Los Rios

My friend Linda Magee had moved to Los Rios and I wanted so much to move there so we could spend more time together. The Housing Division would post the houses available at the Balboa Post Office. You were chosen by your years of service and your grade. Every chance I got I would look for a house in Los Rios. Finally one came up and I convinced my dad to put in for it, and we got it. In 1977 we moved to Los Rios and my parents lived there until my dad retired from Marine Traffic Control in 1991.

We were so happy to live in Los Rios even though the house was a two family home. Paul William attended Los Rios Elementary School, which was right behind our house. I could walk to Linda's house that was right around the corner. We walked everywhere throughout Los Rios and made many friends. We all went to Balboa High School and have many fond memories of all our adventures together.

After my dad retired my parents moved to Panama City where they still live, so we visit often. Every time any of us go to Panama, our families get the tour of the houses we lived in. My 20 year old always says "mom, I know you used to live here, we saw it last time we came." Even though the outside of the houses have change and different people live in them, or they no longer exist like 0944 Amador Road, the houses I lived in made me who I am.
- Helen Hurst Loera

Los Rios: a Flood of Childhood Memories

The words, "Los Rios," bring back a flood of terrific childhood memories: Los Rios Elementary, catching a chiva to Balboa for a dime, riding our bikes with balloons clipped to the spokes with clothespins, Christmas lights decorating houses year-round, and white hibiscus bushes dotted with soft red blossoms. We could meet our friends for Double Dutch jump rope, touch little "shame brier" plants to make them close, or go to Fong Sick's Chinese garden for Chinese plums or ice cream beans. The neighborhood was full of musical sounds: the passing train whistle, the sing-song call of the Pepsi man, the flute scale of the knife sharpener, transistor radios, SCN on the television. For entertainment we took the train to the Morgan Gardens fair or the Clayton fair, looked for bats hanging under the roof overhangs, and rode our bikes down to the animal hospital at nearby Corozal. Once our big Amazon parrot, Popeye, flew away to the Los Rios Elementary school and landed in the top of the huge mimosa tree. My dad had to climb way up to get him down while my uncle gave directions from below. We bought our Christmas trees at Section I, the biggest building in the world, it seemed, and decorated them with big old fashioned lights, bubble lights, and armfuls of silver icicles. We lived next door to Harned and Dunn, the dance teachers, and our big collie drank out of the huge fountain in their front yard. Los Rios days were good.
- Melissa Ward Forney

Playing in Los Rios

In Los Rios we used to have bonfires. I was around nine years old then. After Christmas, we'd usually have our bonfires around January third or fourth. We'd get the big gangs together, and we'd all contribute Christmas trees; we'd collect about 1,500 trees at the most. It wasn't really the parents involved; it was the kids, but the parents offered transportation.

If anybody got caught ripping off the trees, they'd get punched out. We kept the trees at certain people's houses. We'd pile them on the roofs, about three or four layers high. One time I kept about 60 trees in my house. One night we went out and when we got back, every single one was stolen. I found out who had stolen them and shot my BB gun at them. They had the trees stashed up on Tank Hill. I recognized my trees because I had marked a few of them.

About five o'clock we'd unstack the trees and line them up in the road. Then we'd get the trucks together and take the trees out to Farfan Beach for the bonfire at low tide. After they burned, the tide came in and washed them out. We didn't let any intruders in. You'd have to pay a toll of ten or more trees. There would be 50 to 100 kids there, maybe their parents. Everyone contributed, and if you were a brother or sister you could come; but if you were anybody else, you had to contribute. We did this almost every Christmas. Once in awhile just my family would collect 30 or 40 trees and burn them ourselves. We'd take a case of beer and a case of coke. Dad would drink beer; I'd drink coke. Then we'd go eat somewhere. We weren't mean unless you got caught doing something wrong; then it was the short end of the stick. We weren't really a gang; we were just a Christmas gang. Mostly, we were just friends. Once my friend and I went to Clayton looking for trees. We went by three barracks full of Humungous trees in the garbage. We tried to haul them on our bikes out to Los Rios, but they were at least ten feet long and weighed over 40 pounds. We dragged them about half a mile and stashed them so we could come back the next day.

We also went butt sliding and bike crashing a lot, over by the Administration Building. We used to call it the Miniskirt. We'd ride our bikes down the hill and take a jump, which made us fly about eight feet in the air and wreck into palm leaves.

There was a ditch we used to get Porky to pull us on. The ditch was slimy during rainy season, and he'd get a rope and line up about five people in the ditch. He ran and pulled, and we'd go shooting down the ditch. We had to keep our elbows ut of the way because there were a lot of pipes sticking out of the sides of the ditch. We slid in the monument when it was empty. We'd crash out on skateboards or bikes. We got a big box once and fit six people in it. We topped it and started down the hill. We hit a rock at a fast speed, the box flipped over, and everyone tumbled out. There were big rocks sticking out of the ground. It was good in rainy season and dry season, too, when the grass was kind of fluffed up. We slid on our stomachs sometimes, too.
- Paul O'Donnell in 1978 for a folklore class taught by Mary Knapp at Balboa High School

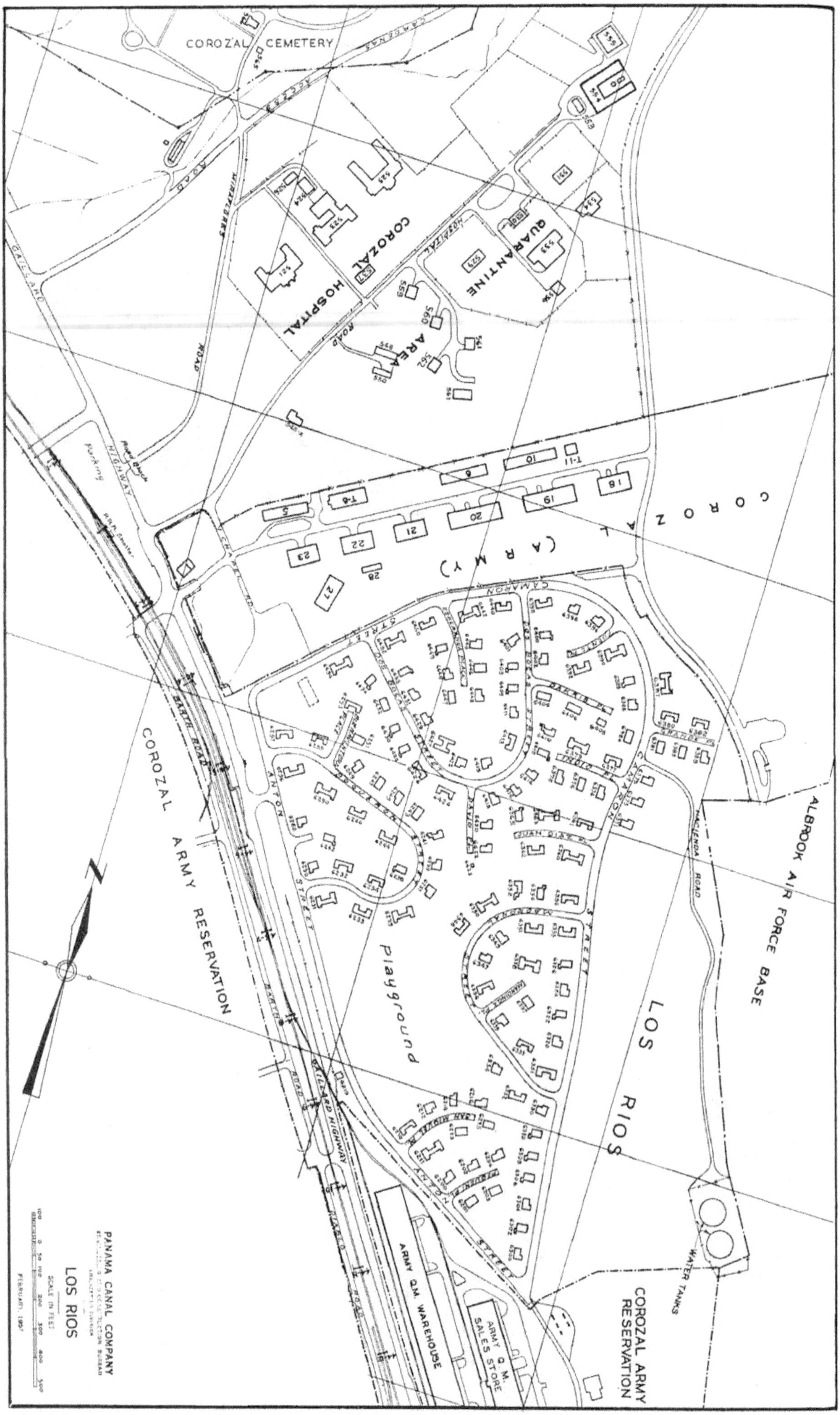

COROZAL CEMETERY

COROZAL HOSPITAL

QUARANTINE AREA

COROZAL (ARMY)

COROZAL

ALBROOK AIR FORCE BASE

LOS RIOS

COROZAL ARMY RESERVATION

Playground

CAMARON STREET

CANON STREET

ANTON STREET

BOYD WAY

GAILLARD HIGHWAY

SANTA ROAD

MELENDA ROAD

WATER TANKS

COROZAL ARMY RESERVATION

ARMY Q.M. WAREHOUSE

ARMY Q.M. SALES STORE

PANAMA CANAL COMPANY
ENGINEERING & CONSTRUCTION BUREAU
SPECIAL ENGINEERING DIVISION

LOS RIOS

FEBRUARY, 1957

SCALE IN FEET
100 0 50 100   200   300   400   500

Margarita

Sponsors

Lester, "Chefa," Gladys & Lena Barrows
The Joseph & Carol Coffin Family
Louie, Barbara & Jon Dedeaux
Harry, Mary, Katherine & Billy Egolf
George, Bobbi, Bruce & Debra Egolf
Andy Nash English
Dennis & Peggy (Hale) Huff
Joan McCullough Ohman
Virginia Kleefkens Rankin
Michael & Elaine Stephenson Family
Family of James & Stacia Walsh
Joseph H. White Jr., & Family

Your Town - Margarita

The Panama Canal Review, December 3, 1954, pp. 8-9, 12.

MARGARITA today is a far cry from the town it was a decade ago. Shrubs and flowers now surround many of the attractive quarters in the newly-developed sections of the townsite.

A little over 15 years ago Margarita was only an expanse of rolling hills; unlike most Canal Zone communities its locale had not occupied even the most minute niche in Isthmian history.

Today Margarita is potentially the Atlantic side's major town, which may outpace, in population and importance, history-laden Cristobal. A year ago Governor Seybold told Civic Council representatives that he foresees Margarita eventually as the Balboa of the Atlantic side, under the growing conception of the Canal Zone as two large urban communities.

Gradually Margarita is being developed to meet this conception. The long-desired swimming pool is now an item in a future budget; a deposit library was opened last summer as a branch of the Canal Zone Library; Margarita is the only Canal Zone community with two elementary schools; and other changes to make the town into a modern major community are still in the long-range planning stage.

Named For Island

Indirectly Margarita derived its name from the little island which is now Fort Randolph, but where that island, originally known as Margarita, got its name is lost in history.

In 1917 a concrete road was built from Fort Randolph to Mount Hope. People had not yet gotten used to the new name of Fort Randolph and the highway was commonly known as the Margarita Road rather than as the Randolph Road.

That same year the Commissary Division established a hog farm "in the Mount Hope district on a point on the new Margarita Road;" the farm, quite naturally, was known as the Margarita Hog Farm and its location is where Margarita now stands.

The road to the Hog Farm led off Diversion Road about where present 5th Street runs; for years after the farm was abandoned (after an outbreak of hog cholera in the late 1920's) this general area was one of the most popular Lovers' Lanes on the Atlantic side.

During the decade between 1930 and 1940 a Panama Railroad conductor, G. G. Boynton--whose hobby was hunting--used one of the farm's old buildings as a kennel for his dogs. His lease was canceled in January 1939 when the Canal's Third Locks began to emerge from the planning stage into a more imminent reality.

Agua Clara Too Hilly

Since the Third Lock's largest single project was to be the triple flight at Gatun, it was obvious that there would have to be a settlement reasonably close by to house the construction forces. Gatun itself was not suitable but if the terrain around Agua Clara had been more to the engineers' liking, Margarita might never have been developed.

It was not until April 1940 that the "Mount Hope area"--in brackets was added: [Margarita' Hog Farm]--was recommended for the Atlantic side's new townsite. It was more suitable, the planners felt, than Agua Clara because it was "oriented for the breeze, better adapted for road grades and building sites, with more space for garages and recreational areas." Furthermore, it would cost less to develop and be "more suitable for a permanent future town."

Four months later Margarita's first buildings were authorized; by Christmastime there were five families living in Margarita. Margarita's first resident was C. E. Borgis, a locomotive crane operator with the Municipal Engineering Division. The apartment into which he and his family moved on Christmas Eve 1940, is now occupied by Miss Mary L. Mehl, a second grade teacher at South Margarita School

Firemen And Policemen

Fire and police protection for the vast array of construction equipment and material which was stacked everywhere was a "must," so into Margarita's first four-family house moved two policemen, Gaddis Wall and Isaiah A. MacKenzie, and two firemen, Lt. W. E. Jones and E. L. Cotton. Captain Wall now is in charge of Cristobal's detective force, Sergeant MacKenzie is retired, and Captain Jones heads Balboa's fire district to which Lieutenant Cotton is assigned.

They set out to make other Margarita pioneers feel welcome. Until the clubhouse kitchen was ready, bachelor Margaritans "messed" at the fire station. Behind the police station, on a little hill, whitewashed stone letters bade newcomers: "Welcome to Margarita, C. Z. Police."

As Lieutenant Cotton recalls it, the first Margaritans were a "good-natured bunch" who made the best of the mud and the board-walks, construction noises, and long hours. Everything but the sandflies. They were the plague of Margarita and not to be taken lightly. Those first residents bought pyrethrum in 10-pound bags and burned it in their houses, in bachelor quarters, and in such public buildings as Margarita had.

Eventually, as the town grew and new insecticides were developed, sandflies became less of a pest but Margaritans today, somewhat immune, still see an occasional outlander guest slapping surreptitiously at legs or ankles.

PALM trees, which border Margarita's civic center, were only a few feet tall 10 years ago. The post office is on the far side of the parking lot; the commissary and service center face two other sides, and 12-family quarters, the first to be occupied in Margarita, are on the fourth side.

FRANCES MOOMAW, Principal of South Margarita School

LEALAND LARRISON is Postmaster for Margarita

HELEN RUSHING, Principal of North Margarita School

Town's First Party

By March 1941, Margarita was ready for its first community affair; the location was the newly completed fire station. Some 200 Margaritans and friends turned out for the party; it was supposed to last from 7 to 9 p.m., but was still going strong at 3 a. m.

Margarita, in its early days, was a town divided by employers—spiritually and physically. In one section, around the present community center, were the homes of Third Locks people and a handful of others like the policemen, firemen, clubhouse, and commissary employees.

A short distance away, where the Church of the Holy Family and the Knights of Columbus Hall now stand, were the homes of the contractors' employees. They lived, for the most part, in tiny prefabricated houses which everyone called "doll-houses."

"I never saw anything grow like that Contractors' area," Captain Wall says now. "For a while the doll-houses were popping up faster than one a day."

Years later some of the doll-houses were sold to the Panama Cement Company, they are still standing in the company's little town just off the Boyd-Roosevelt Highway near the cement plant.

Fast-Growing Town

Margarita grew fast, once it started. From a half a dozen families in December 1940, the population increased to 1,032, a fifth of them children, in 1943. Margarita had its own hospital, only recently

demolished, quarters for the hospital staff, a commissary, clubhouse, post office, gymnasium, elementary school and kindergarten. For years all construction halted at Espave-named for one of Panama's largest forest trees which abound in this area-and 3d Streets.

The exception was Ghost Hill on the high ground not far from Fort Gulick where Lychee Street was built later. It was a smallish clearing, surrounded by jungle, with a dozen or so houses, and a lonely place to be on foot patrol at night, Captain Wall remembers.

Although Margarita was definitely a construction town, it was fairly well-behaved. There were incidents, to be sure, and some funny and unprintable ones, but by and large there was remarkably little of people poking other people in the nose or bottle parties which went on to all hours.

At first Margaritans had to go "to town"-Cristobal or Colon-for their fun. Some of them belonged to the Progressive Dance Club which met regularly at the Hotel Washington. Then came the still-flourishing Margarita Recreation Association which was the subject of a REVIEW story in June 1953 and which has sponsored everything from dances to picnics to hobby groups to Scout Shacks. It once even had its own weekly newspaper, whose slogan was: "All the news that fits we print."

With the cessation of the Third Locks project, old Margaritans say, the town "practically died." The contractors people packed up and went away; the Third Locks force was cut to a clean-up squad. But Margarita's population in 1944 was 854, which doesn't sound very moribund.

It wasn't long before Atlantic siders began to see the advantages of suburban living; despite gasoline rationing a shift in population from Cristobal to Margarita began during the last of the war years. This was spurred still further in 1945 when 11 new two-story, two-family, houses were built in the Casuarina and Hevea Place neighborhood.

New Quarters

When Margarita got its first on-the-ground masonry houses in 1948, close to a thousand people turned out to inspect them. The Chief Quartermaster, then the Grand Mogul of Housing, later wrote their occupants asking what they thought of the new houses. Their replies were highly

complimentary.

In 1950 Margarita began to boom. Barracks which had housed Third Locks bachelors were demolished to make room for Margarita's re-development. The townsite was extended and new houses began to appear. The first of these were occupied in October 1951.

Today, Margarita is bounded, more or less, by a horseshoe made up of Espave and Margarita Avenues, on the east and west. Its numbered streets run, also more or less, north and south.

Margarita, which has little past, is a town with a future. It has the most modern elementary school building in the Canal Zone. Two churches serve Margarita and land has been assigned for three others. The Elks and the Knights of Columbus have their own buildings, as do the Veterans of Foreign Wars. The Atlantic Side Saddle Club, the Brazos Brook Country Club, the Cristobal Gun Club, and the Colon Humane Society's kennels are practically next door.

Its Civic Council, consolidated with that of Cristobal, is an ardent advocate of Margarita's progress. One of its latest projects is a Teenage Club, in a building recently assigned by the Governor. It is being refurbished by the youngsters and some parents and will be run by the teenagers, with a Civic Council committee standing ready to give advice, when asked.

Margarita's town Halloween parties are famous and its Fourth of July celebration serves the people of the

G. J. MARCEAU, Manager of the Margarita Service Center

E. T. HARPER, Manager of the Margarita Commissary

Atlantic side. Its town spirit has kept pace
with its growth. One old-time Margaritan
has a peculiar thermometer for both.

"See those palm trees," he said the
other day, "they were not over a couple of
feet high when I came here 14 years ago.
Now some of them are topping 30 feet.
That's the way Margarita has grown too."

Sgt. STARFORD L. CHURCHILL and Lt.
JAMES BARTLETT head Margarita's
fire-fighting forces.

Program for Canal Zone Girl Scouting Covers Range
of Year-Round Activities

The Panama Canal Review, July 4, 1952, pp. 2, 15.

Although there were Girl Scouts
during the 1920's, the Scout Council
was not formed
until 1934 with the first troops at Fort
Amador and the Post of Corozal.
Today every community where there
are girls of scouting age is represented
in the Girl Scout program. Every troop
is the girls' own club, for which she
works and plans. And through her program activities she learns what skills
and talents she may have and how to
use them, how she may be of service to
her community and how she can foster
international friendship.

Pictured here on the Atlantic side
is Troop 38, Margarita, C. Z.

Margarita School

The Panama Canal Review, September 4, 1953, p. 6, and January 1, 1954, p. 7.

The greatest change in schools
this year is the opening of the new elementary school plant at Margarita,
the South Margarita School, which
has been under construction since last
March. The old elementary school at
Margarita, which will continue in operation, will be known as the North Margarita School. The new school has 15
classrooms in four parallel units with
a fifth unit for administrative use. The
new school is constructed on a modern
type plan new to Canal Zone school
buildings.

Margarita's new elementary school, August 1953, still under construction

The Panama Canal Review, September 4, 1953, p. 6.

Sixty children from the fourth, fifth, and sixth grades of the new South Margarita School caroled gaily in December as the new building was formally dedicated. Miss Odell Waters was the music director. Nearly 800 parents and other visitors attended the inauguration which was held in the patio of the school building. The formal ceremony during which the keys to the school were presented to S.E. Esser, Superintendent of Schools, by Lt. Gov. Harry O. Paxton, was preceded by an open house in all of the classrooms for the children's parents.

The school is located in a nine acre area about midway between the older and newer sections of Margarita. It consists of five one story buildings, connected by covered passageways. Its classroom walls are painted in pastel colors and the floors are of harmonizing asphalt tile. All furniture is new. Plants have been planted in the patio. The school is expected to be one of the beauty spots of the Canal Zone.

AIR CONDITIONING COMING TO THE ATLANTIC SIDE

The Panama Canal Review, September 6, 1957, p. 11.

As the conversion of 25 cycle current to 60 cycle current moves south from Cristobal, more and more Zonians are relaxing in cool, dehumidified air to the soothing hum of one or more air-conditioning units. Air conditioning of Zonian homes provides not only comfort, but also for children with asthma the difference between sickness and well-being.

In Margarita, between 40 and 50 homes are air conditioned, entirely or in part. Several householders have fitted their dwellings with two units and have the entire home air conditioned. Others have limited themselves to air conditioning their bedrooms.

Mr. and Mrs. S.A. Hammond, of Margarita have a one-ton unit, which can make about 600 square feet of floor space comfortable. A second unit, one of the three-quarter ton type, will be installed soon.

Air conditioners are not being given away, but a unit costs less than some TV sets. Three quarter ton units can be had from $207 to about $230, ordered through the Commissary Division, and one-ton units from the same source range from $240 up to about $275.

"Makes us really comfortable." says Mr. S. A. Hammond, Margarita.

Two family duplexes on Sixth Street.
Courtesy of Dick Cunningham.

Four family two bedroom house on circle of Heavea Place. Note how a resident enclosed the carport.
Courtesy of Dick Cunningham.

A two family duplex on Heavea Place. Note the fence around the house and the columns on one side.
Courtesy of Dick Cunningham.

Cottage on Campana Place.
Courtesy of Joan Ohman.

Renovated two-family house with an added porch on Hevea Place, circa 2011.
Courtesy of Joan Ohman.

Memories

Margarita Hospital: A Get Rich Quick Scheme

This is a story my husband told me about his childhood growing up in Margarita.

I am sure everyone remembers the old Margarita Hospital. Well after it was deserted and before it was torn down, Billy Rankin and his buddy, Butch Tobin, thought of a way to get rich quick. They knew all the windows in the hospital were the type where lead weights were used as pulleys in the windows to make them go up and down. Billy and Butch got their wagon and went over to the old building and proceeded to remove all the lead weights from the windows. They hauled several wagon loads up to the hill where the Elks club was later built and dug a deep hole and buried all these weights, to be retrieved at a later date and to sell the lead for cash. Before they could get around to retrieving their bootie, the Elks club was built on that hill and to this day, there may still be a pile of lead weights buried beneath the building.
- Virginia Rankin

Margarita Tree Burn Activity

Everyone remembers the Christmas tree burns and how gangs collected the trees to see which gang had the most trees to throw on the bonfire. Well, those of us who lived on Snob Hill in Margarita had all our trees hidden in the maid's room of Tabers' house. Needless to say, Jack Taber, who was the fire chief, didn't like that much. Anyway, on Saturday afternoon, we kids from Snob Hill went to the Margarita Clubhouse to attend the matinee. When we got back after the movie, we saw all our trees were gone and there was a needle trail from the Tabers' house down the hill side and through the jungle to Sixth Street and it ended up at Billy Rankin's house. His gang had taken advantage of no guard at our stash of trees to liberate them to add to their Sixth Street pile.

Several years later, Billy's mom got a phone call from some irate parent complaining of all their children's trees being stolen and as Billy, who was older now and out of the tree stealing business, came walking into the room, he heard his mother telling this parent, "It couldn't be Billy who took the trees. He is retired from the business."
- Virginia Rankin

Christmas tree burning. *Courtesy of Lesley Hendricks.*

My Canal Zone Home - Margarita

In December of 1968, my dad put in for a promotion with Marine Traffic Control and we moved to Margarita. We first lived in a small 2 family house and within a few months moved to a larger 2 family home where we lived for 1 and ½ years. I loved living in Margarita, I remember getting my B badge and going down to the pool with all the neighborhood kids. I went to Playville Nursery School while living there. Right behind our huge house was a hill and at the bottom a large field. Paul and I would run around with nets trying to catch butterflies!
- Helen Hurst Loera

South Margarita Elementary School in 1953, the year it opened. *From the Panama Canal Museum Collection.*

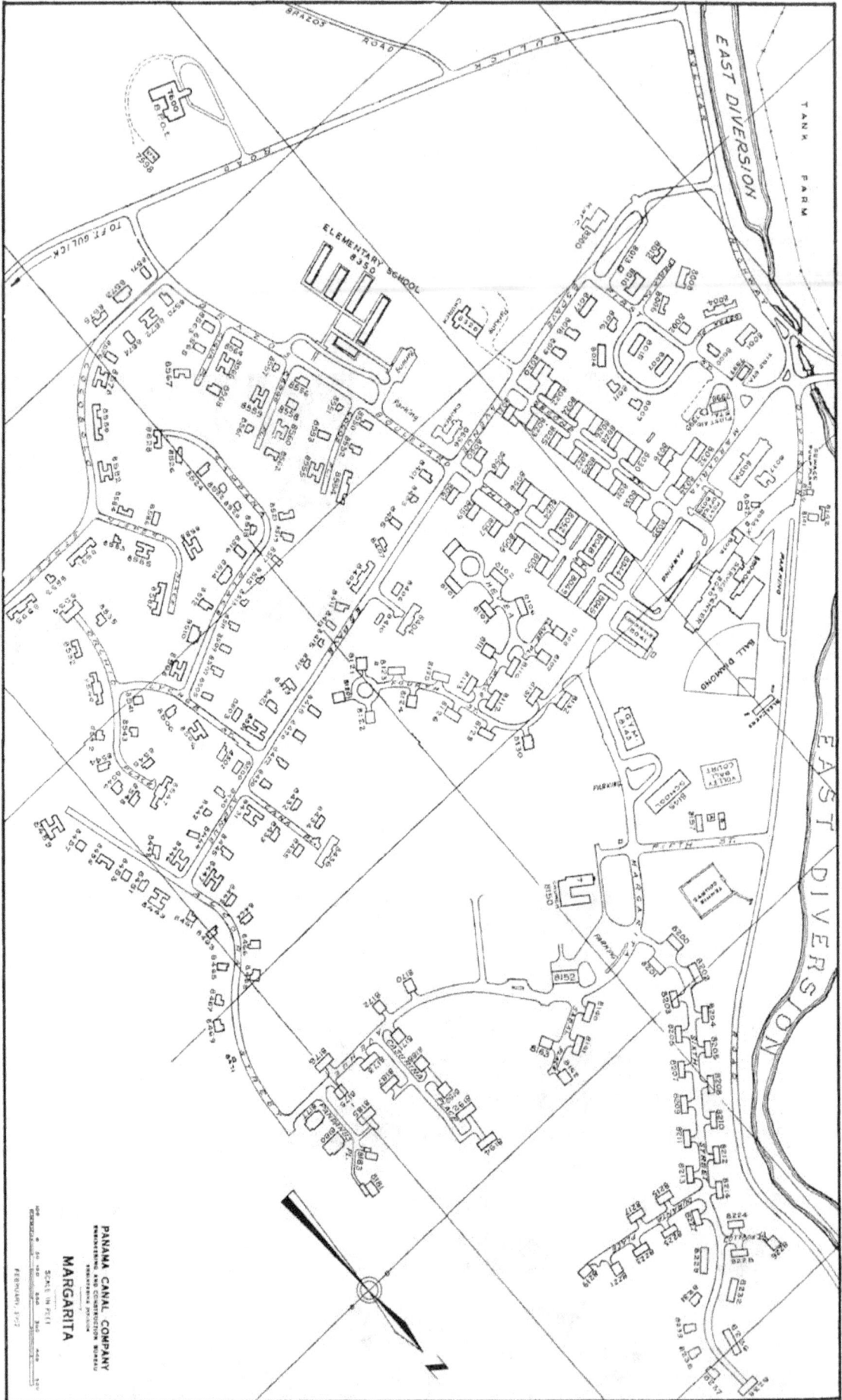

MARGARITA

PANAMA CANAL COMPANY
ENGINEERING AND CONSTRUCTION BUREAU
ENGINEERING DIVISION

SCALE IN FEET

FEBRUARY, 1972

EAST DIVERSION

TANK FARM

BRAZOS ROAD

TO FT. GULICK

GULICK

ELEMENTARY SCHOOL
8350

BALL DIAMOND

VOLLEY BALL COURT

PARKING

FIFTH ST.

114

Paraiso

Sponsor

In Memory and Honor of the Residents of Paraiso

Your Town - Paraiso

The Panama Canal Review, November 4, 1955, pp. 10-12.

PARAISO is still the graceful undulating valley described a century ago; the section in the foreground is Beverly Hills, with Dogpatch just below. From Beverly Hills to the lower end of Paraiso near the ballpark is just a little less than a mile.

Paraiso isn't quite like the mythical phoenix which had the ability to resurrect itself, and quite frequently, too, from its own ashes. Paraiso was never reduced to ashes but it has had more incarnations than any community in the Canal Zone.

In the days of the buccaneers, it was a stop on the "dry-season trail" between the Atlantic and Pacific; early Canal Zone fable had it that Sir Henry Morgan first saw Old Panama from a hilltop near Paraiso. Whether he did or not is important only to historians, but the tower at the old city can be seen plainly on a clear day from the crest of the hill across Gaillard Highway from Paraiso.

And in the past hundred years, Paraiso has been: A station on the Panama Railroad; the headquarters of a chantier, or working section, for the French Canal Company; an American construction day town; Dredging Division headquarters; a colored community; an Army camp; and, today, the largest local-rate town south of the Canal Zone's continental divide.

Beautiful Paradise

During the 1850's, when surveyors and engineers were laying out the railroad line, they found a pass which led into what Otis, a few years later, described as "the beautiful undulating valley of Paraiso, or Paradise, surrounded by high conical hills where Nature in wierd profusion seems to have expended her choicest wealth." The railroad tracks were laid in a 40-foot deep cut; slides brought on by heavy rains once covered them with 20 feet of earth and rock, and weeks of work went into clearing the rails again.

Paraiso itself, eight miles from Panama City, was the first Pacific coast stop after the trains had passed "The Summit." During the mid-nineteenth century, it was little more than a native settlement, although a fine natural spring gave it more importance than most way stops.

Toward the end of the 1800's, Paraiso became important for the first time. It was a key spot in all the French Canal plans and was repeatedly selected as a site for one or more of the locks. When the first French Company began work in 1882, Paraiso was the southernmost spot where dry excavation was carried on.

Years went on, the French encountered troubles, but the work at Paraiso continued. In 1888 the Star & Herald reported that "large and heavy trains of Decauville dumping cars" were hauling load after load "out of the work and up a steep incline." Work had been started to relocate the railroad clear of the cut and a bridge was being built on which the railroad would cross the Canal.

Eventually, the French construction became little more than a token effort and its force dwindled away; at the turn of the century Paraiso's population was about 800, living in "125 frame houses and 100 huts." When the United States bought the rights and properties of the French Company in 1904, some of the Paraiso buildings were still usable. Among them was a 10-room office building, a two-story affair with five rooms on each floor. When carpenters and painters began to recondition it, they found

ELLIS L. FAWCETT
President, Paraiso Civic Council

in one room records of the French Company, letter presses (small machines used for copying letters), a safe, and maps of Paraiso and Pedro Miguel. Another French building which became American quarters was a two-room mission house which, when the Americans came, still had its church bell over the door.

The French Company had had small machine shops at Paraiso. These the American forces enlarged by moving some buildings from Culebra and adding a shed where locomotives were hostled. Light repairs were done at the Paraiso shops to all kinds of equipment at the southern end of the Canal Zone; heavier repairs were made at Gorgona or Empire.

In 1908, when the Canal work was reorganized, the Paraiso shops were abandoned, the 265 men who had worked there were distributed to Empire and Gorgona or elsewhere, the buildings themselves were used for storehouses for material for Pedro Miguel locks, and Paraiso became more or less a residential community for the locks forces or for the railroad engineers who worked out of Pedro Miguel.

Construction Days

A good many Isthmians still remember Paraiso in those early days. J. J. McGuigan, once chief clerk of the Sanitary Service of the Canal Zone and much more recently District Attorney for the Canal Zone, lived in Paraiso from 1906 to 1908 when it was headquarters for the Chief Sanitary Inspector, J. A. LePrince.

In 1906, he recalls, Paraiso's railroad station was on the west side of the Canal excavation. A passenger from Paraiso had to make his way down one bank, across the almost inevitably muddy flat which later became the bottom of the Canal, and up another steep embankment. Most people stayed home or took a train at Pedro Miguel station.

Mrs. Dorothy Hamlin of the Accounting Division was only a small girl when she and her family, the Charles Magnusons, moved to Paraiso in 1910. She remembers getting off the train at Pedro Miguel and boarding a "brake" for the ride to Paraiso. The conveyance was drawn by two mules and driven by an old colored man named Dixon. For years he and his mule team were Paraiso's taxi.

As in all Canal towns, the size and quality of quarters depended on the salary of the family head; locomotive engineers, who drew good salaries, were the aristocracy of Paraiso and lived in the town's better houses. Bernard McIntyre, now the Panama Railroad's senior engineer, was the son of one of Paraiso's elite.

Slides and Socials

Construction-day Paraiso lay right on the edge of the Canal excavation. Every once in a while the unstable banks slid into the cut, houses and offices were hurriedly vacated and their occupants moved to safer spots farther from the edge. In 1908 slides necessitated the removal of a great block of the "native" quarters.

The commissary and post office were close to the excavation; the two-story lodge hall used on Sundays for church services and the rest of the week for social or fraternal activities were farther inland. There was a hotel—which today would be called a bachelor mess, a 16-bed hospital, one of the half dozen bandstands in the Canal Zone, and a public market which handled fresh vegetables and fruits to supplement the commissary's supply.

Paraiso's social life was pretty well self-contained. It had a Woman's Club, organized in 1907; its original president was Mrs. J. C. Barnett, one of the first women to make her home in the Canal Zone. A Dancing Club held practice sessions every Monday night and dances on Saturday; there were chapters of the Eagles, Kangaroos, Red Men, and Sojourners at Paraiso. A unique organization was the Texas Whist Club, whose members were married couples who had purchased property in southern Texas with the idea of making their eventual homes there.

Paraiso has played host to two Presidents. Early one morning in November 1906, a construction train shunted down through the Cut to Paraiso, bearing President Theodore Roosevelt—the first President to leave the United States during his term of office. According to Mr. McGuigan, "Teddy made one of his short, forceful talks to the workers gathered there, complimenting them for being on the job notwithstanding the heavy rain that was falling and the muddy ground underfoot." Paraiso's second Presidential guest was President William H. Taft who gave an address at Paraiso in 1910 when he was on his fifth visit to the Isthmus.

Canal Crossing

During the construction period the railroad crossed the canal channel at Paraiso on a 500-foot wooden trestle. As construction progressed this was replaced by another bridge closer to the locks. After the Canal was finished and the railroad located entirely on the east side of the Canal, a pontoon bridge, whose base was built at Mount Hope and towed through the Canal to Paraiso where the superstructure was erected, provided access to the west bank. Before the idea of a pontoon bridge was adopted, a tunnel under the Canal at Paraiso had been considered, and the idea discarded because of the cost.

In July 1913, Paraiso was selected for headquarters of the Dredging Division, the organization which succeeded the Sixth Division of the construction period. The old machine shop was refitted to replace the shops at either end of the Canal where the dredges had been repaired up to that time. From this base at Paraiso, dredges worked

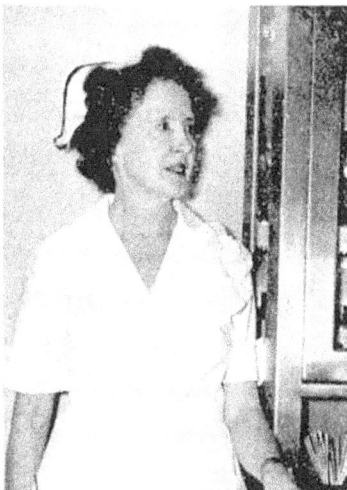

Miss MARIE V. BRAUER
Nurse, Paraiso First Aid Station

day and night during the period soon after the Canal was opened when slides repeatedly blocked and eventually closed the waterway for a nine-month stretch.

About the time the Dredging Division moved into Paraiso, one of the town's oldest activities moved out. This was the Coca-Cola Bottling Plant, which had set up operations in 1905 on the west bank of the Canal opposite Paraiso. It had been started by W. N. Seitz, who operated it mainly as a soda water factory until it was sold to the Panama Coca-Cola Company. The Paraiso site had been chosen because of its proximity to the Paraiso springs which shared with the springs on Taboga Island the reputation of being the purest water in this part of Panama.

These same springs had for years supplied drinking water for Corozal and other construction day towns. A special train each day hauled dozens of demijohns of Paraiso water to the towns along the line.

"Silver" Town

By 1918 the danger from slides had abated and the Dredging Division's force was reduced. That year, Paraiso's American families were moved to Pedro Miguel and their old quarters, together with the one-time hotel and other buildings were converted to homes for "silver" families.

For the next decade or so, Paraiso was an undistinguished, run-of-the-mill, Canal settlement. The quarters were grouped in little subdivisions known as Jamaica Town, Hamilton Hill, and Spanish

NOLAN A. BISSELL
Postmaster in Paraiso's Post Office

Town. The Paraiso school, however, was outstanding. For years its principal was the Rev. D. A. Osborne, known far and wide as "Teacher Osborne." His son, Alfred E. Osborne, is today Supervisor of Instruction for the Canal Zone's Latin American Schools. The building itself was the first modern school in Pacific side colored communities; its school garden was the pride of the town; and the quality of its teaching was borne out by the number of Paraiso students who went on to the Normal School to become Canal Zone teachers.

During the latter part of the 1930's, headquarters of the Dredging Division moved to Gamboa and Paraiso went out of existence entirely as a Canal Zone community. In November 1939, just a year after the Canal abandoned it, Paraiso became a military post.

Camp Paraiso

Troops of the Fifth Infantry moved in, built barracks and quarters, a movie theater, and a post exchange. Bayonets bristled in the "beautiful undulating valley of Paradise." For a while, the Army postal locator unit was at Camp Paraiso. Close beside Paraiso what was known as a killer net stretched to a hilltop across the Canal to trap unwary dive bombers. But as war moved farther from the Canal Zone Paraiso's forces began to dwindle and in 1943 it was closed as a military camp.

The following year Paraiso again became a Canal Zone town although for a time the Army retained a few buildings at the upper edge of the old camp. Quarters built by the Army were remodeled into family residences and barracks became bachelor quarters. The Army theater and post exchange became a clubhouse, the commissary was reopened, and a new school, now the Junior High School, was built for the elementary grades.

Modern Paraiso

Today Paraiso is one of the most modern of the Canal Zone's communities. Its elementary school, for two years an Isthmian showpiece, is now overshadowed by the new Civic Center which was formally opened last month. Aside from being the first building designed as a civic center—it houses a first-aid station, post office, a luncheonette, and a meeting room, it is also the first canal building through whose roof trees grow on purpose. Future plans for Paraiso call for a new commissary adjacent to the civic center, and a new high school will open there next year.

R.G. ROWE
Manager, Paraiso Commissary

C.C. GITTENS
Acting Manager, Paraiso Luncheonette

PARAISO 50 years ago lay close along the excavation for the Canal. From time to time, as the banks crumbled, buildings had to be moved out of danger. In the middleground, toward the right, is the Lodge Hall with the bandstand close by; the commissary and post office were on the edge of the excavation on the right, but out of sight here.

Last fall's census gave Paraiso's population as 3,008, just about that of the construction period. Paraiso families live, for the most part, in modern, concrete, two-family houses, the first of which were occupied in February 1953. Paraisanos have their own names for the various parts of their town. Lakeview is near the Canal; Spanish Town along Gaillard Highway; Ghost Town, naturally, close to the cemetery—where the graves of two French engineers are evidences of days gone by; Beverly Hills is the "heights" of Paraiso; and Dogpatch lies in the little depression just below Beverly Hills. Dogpatch, incidentally, belies its comic strip name. It is one of the neatest and best-landscaped sections of Paraiso and its residents take special pride in their outdoor Christmas decorations.

Paraiso's school enrollment – children-wise – is 818; its elementary school is the second largest of those in the local-rate towns. But Paraiso's elders also go to school. Two groups are taking night classes in Spanish, under the sponsorship of the Civic Council, one of Paraiso's most active organizations.

Memorial Plaque At Paraiso Honors Korean War Veterans

The Panama Canal Review, December 3, 1954, p. 4.

A memorial plaque honoring three young Isthmians who died during the Korean War and commemorating all of those from the Canal Zone and Panama who served with the United States Armed Forces during the Korean conflict was dedicated November 11 at Paraiso.

The memorial, which is located at the upper entrance to the town of Paraiso, was erected through the joint efforts of the Pacific Army Mothers Club, of which Mrs. Daisy Robinson is president, and of the Mutual Aid Club, whose president is Frank B. Burke.

The United States Army provided a guard of honor while the Guardia Nacional Band represented Panama at the ceremonies.

The servicemen whose names appear on the plaque are Sgt. Jose Molina Ceballos, Pvt. Ben A. Franklin, and Pvt. Gilbert D. Francis.

Beverly Hills, Calif.? No! Beverly Hills, C.Z.

The Panama Canal Review, September 4, 1953, p. 16.

Although the names will never be found on maps, five new communities of just-completed housing in 1953 in the town of Paraiso were suddenly being mentioned in the Canal Zone. The names, the inspiration of residents, were: Beverly Hills, Dogpatch, Lakeview, Jamaicatown and Ghost Town.

Beverly Hills is the hilly section of the town, near the water tower. Dogpatch is the small cluster of new houses at the foot of Beverly Hills. Lakeview is the area nearest the Canal and Jamaicatown the section near the ball diamond. Ghost Town, appropriately, is adjacent to the old Paraiso Cemetery.

OLD PARAISO-- This is how the town of Paraiso looked in 1906, when workers arrived on the isthmus dring the early constrction period. it has twice been abandoned and twice revived during its colorful history.

118

Happy Times in Paraiso - 1942

It was 1942, and Muriel Evans and I were working for the Special Engineering Division, building a new, larger Panama Canal. World War 2 was on, fighting the Japs and Germans, but we were still enjoying life in the Canal Zone, still living like the British in India. I was enjoying meeting the movie stars that often visited. Clark Gable often came to fish in this "land abounding in fish;" that's what the word "Panama" meant in the old Cuna language. I met Jackie Cooper at Pier 18 when Jackie was only 14, Irene Dunne at the old Union Club, William Powell at the Atlas Beer Garden. We were still living la vida wonderful.

One day Muriel said to me, "Please come to a party at my house Saturday night." Muriel lived in Pedro Miguel, right next to Paraiso. She added that a new group of infantry soldiers had just arrived for tropical training, stationed in Paraiso, the 158th Infantry Regiment from Arizona, and she and her sister Barbara wanted to make them feel welcomed. They were known as the Bushmasters.

PARAISO—It did not sound appealing to me. Paraiso dated from French construction days; then our own construction days. President Teddy Roosevelt visited there in 1906. Next for many years only Jamaicans lived there. Now it was Camp Paraiso, APO 833.

Mother would not be happy. We were not allowed to associate with soldiers in the old days. They earned only $15 a month, the same as our maids earned. Paraiso was too far from Balboa where I lived. My grandfather, Richard Roberts, first arrived in 1907 to build the canal; he lived in 760-B, the first concrete quarters ever built in Balboa. He had a Roosevelt Medal. I was a third generation Zonian.

Muriel persevered, said they were the handsomest lieutenants they could find. That did it. I went to the party for soldiers from Paraiso. Thus began the days of happy times.

Fun times continued—dancing in that old wooden building in 1942, meeting all those good looking gentlemen in the 158th. One was Jack Hershey, from Hershey, Pennsylvania. There I met the love of my life, LT Charles S. Stough. Soon we were going steady. It wasn't easy. The commanding officer kept sending him off to jungle training in Camp Chorrera, then deep into the jungle.

Finally, the commanding officer refused permission for LT Stough to marry me! That did it! LT Stough kissed them good by and got reassigned to another outfit in the Canal Zone. By this time the Panama Canal, with lengthy ropes, had huge balloons soaring over all the locks, to keep the Japs from bombing the canal. As the 158th Regiment sailed away to the South Pacific after their year of training, I shall never forget something I read. A 158th soldier looking back, muttered, "They ought to cut all those ropes and let the damned place sink."

LT Stough and I were soon married, so quickly that the photographer failed to arrive.

Thus began a marriage that lasted 50 happy years, producing two fine sons and many fine grandsons. It never would have happened if I had turned down Muriel's invitation. There must be a good reason for naming that town Paraiso, the word meaning PARADISE. Webster's dictionary assures us it also means any place of great beauty and happiness.

- Jeanne Flynn Stough

NEW PARAISO

In 1952, Paraiso was one of the fastest growing communities on the Pacific side. the Canal Zone. A total of 244 two-family units were built by contractor, Tucker McClure along with streets and other facilities. When completed in 1953 the duplexes were assigned to residents of Red Tank.
Panama Canal Review, July 4, 1952.

Paraiso School, circa 1950s.

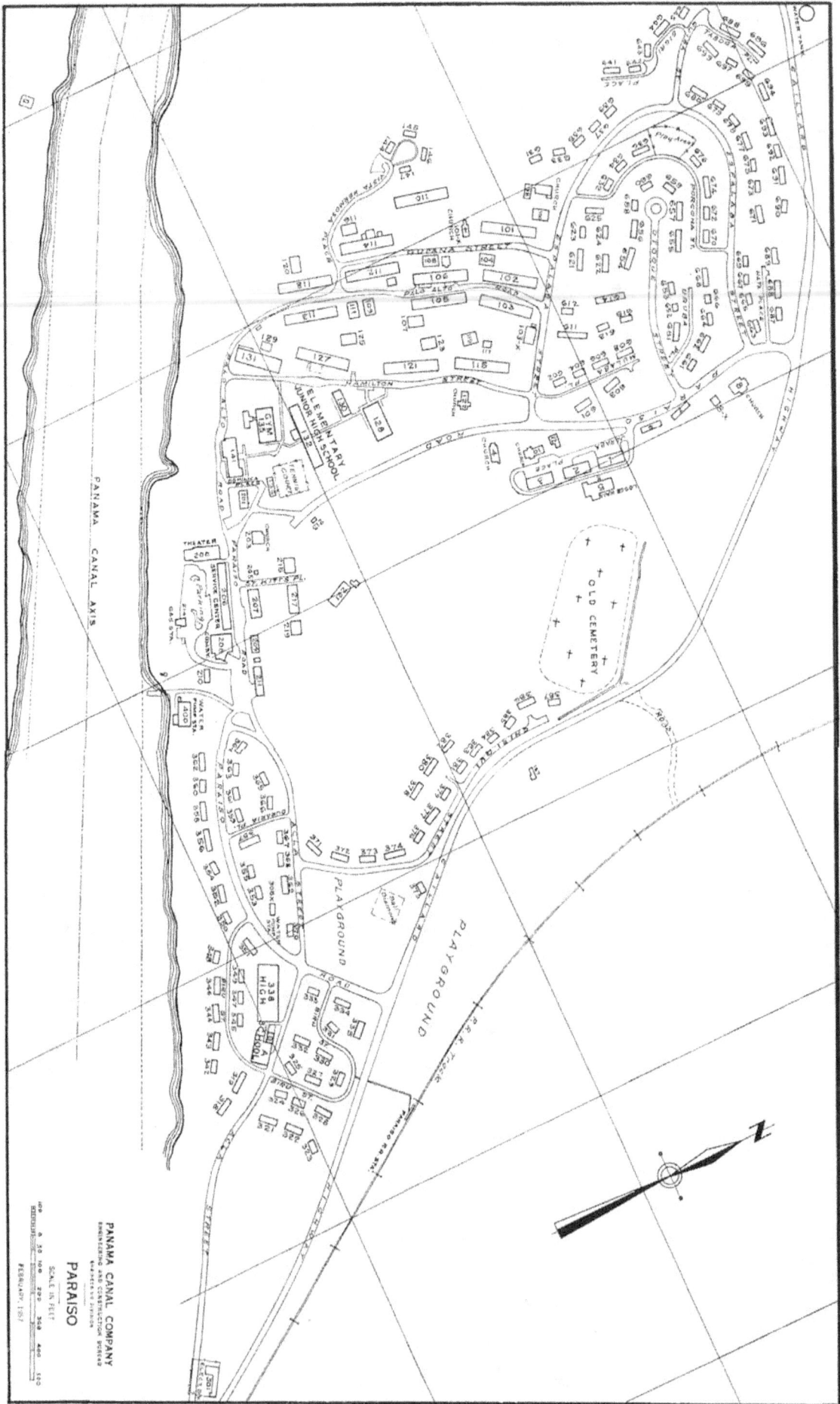

Pedro Miguel

Sponsors

W.R. Dunning Family, Bill, Pat, Danny, Sandi & Vicki
Robert & Delores Leisy, BHS '49
Norma Stillwell Martin
Roy, Geneva & Janet Stockham
The Wards—Marvin, Jackie, Debbie, Lissa

Your Town - Pedro Miguel

The Panama Canal Review, February 5, 1954, pp. 8-10

PEDRO MIGUEL lies between a band of hills and the Locks, which will apprear in this aerial photograph near the top of the picture. In the cleared space in the center, where the ball diamond is now located, Chinese merchants had their shops during the construction days.

Peter Magill it has been for many years and Peter Magill it will probably continue to be as long as there is a Pedro Miguel and Americans here to mispronounce its name. It is even spelled "Peter Magill" in some official records.

The origin of Pedro Miguel's name is a matter of argument among its residents. Adrien Bouche, who has lived there for many years, grew up on the story that Pedro Miguel was the name of a railroad section foreman. There wasn't much of a town in early railroad days so the stop was known as Pedro Miguel's cabin.

Others believe that the town's name is properly San Pedro Miguel—St. Peter Michael—which was the Spanish name for the river. Certainly an 1867 history of the railroad refers to the San Pedro Miguel River, "a narrow tidewater tributary of the Rio Grande," which the railroad crossed on an iron bridge.

Pedro Miguel was off the beaten path of trans-Isthmian travellers of the Spanish colonial days. There was probably a fair sized settlement in the general area; during canal construction days ruins of a large old Spanish church were found not far from Pedro Miguel. Indian artifacts discovered in Pedro Miguel's hills prove even earlier habitation.

French There In 1888

French canal forces started work at Pedro Miguel on January 15, 1888. In May of that year the site, where one lock was to be built, was one-third excavated. Later the New French Canal Company modified the plans and decided that two locks would be built at the Pedro Miguel location.

There could not have been much of a settlement at Pedro Miguel, however, for when the Isthmian Canal Commission began to take stock in 1904-5 of property taken over from the French Company it found at Pedro Miguel nine usable buildings and three in such bad condition they were destroyed.

One of the old French buildings became a police station, another a commissary, and others were used as quarters. Mr. Bouche says that one of these

old quarters is still standing on Miraflores Street near the Boy Scout Shack. He identifies it from its brick underpinnings.

Early in 1905 Chief Engineer John F. Wallace wrote Governor George W. Davis:

"Pedro Miguel is the point, you know, where the principal line from the east side of the south end of Culebra Cut joins the main line of the Panama Railroad and I desire to establish here the headquarters of the men employed in the transportation and excavating work in this general vicinity." (The present road to the Cucaracha signal station is a part of the old railroad bed.)

Kitchens, a messhall, a "hotel," a post office, and bachelor barracks began to rise, all west of the present railroad line. The work was considered so urgent that later that year Governor Davis asked that "work on Corozal quarters be suspended and Pedro Miguel pushed forward."

Railroad Center

Although the ICC selected Pedro Miguel as the site of one set of locks, after a lock-type canal was decided on in 1906, Pedro Miguel remained primarily a railroad center. A nine-track railroad yard, a coaling plant and a repair shop were built that year, and at Pedro Miguel President Theodore Roosevelt made his first stop when he toured the construction line in November 1906. It is pretty certain that all of Pedro Miguel's 754 residents, 79 of them Americans, turned out to welcome him.

Most of the construction force lived at Corozal, if they were Americans, Mr. Bouche recalls, or at "40-mile Camp" which was practically part of Pedro Miguel or at one or another of the nearby "silver" construction camps if they were non-Americans. In 1908, however, the Pedro Miguel mess for European laborers was enlarged to accommodate 450 men; it was the largest in the Canal Zone at the time.

In 1912 John O. Collins described Pedro Miguel: "Here is an engine house where as many as 80 locomotives tie up for the night. One of the most interesting sights on the Canal is watching these locomotives leave the engine house for their work in the morning. The first one leaves about 6:30

a. m. and the last is clear of the yards 10 minutes later."

Pedro Miguel was even then becoming modern. It had had electricity since 1907. In 1908 there were 491 pupils in the Pedro Miguel school; the lone high school student had to travel to Empire for his classes. A commissary, which served all employees, was approximately where a large storehouse now stands on the road to the Boat Club. A volunteer fire company protected the town and a Commission truck garden supplied fresh vegetables. In 1909 there was some excitement when a slide in the locks excavation swallowed up a chicken house and trees belonging to one of the cottages. The house, left only four feet from the slide, was moved hurriedly.

Firsthand Report

For a description of Pedro Miguel life in the late construction days, THE REVIEW turned to Mrs. Eula J. Ewing, who lived in Pedro Miguel from 1911 until her retirement in 1952. For many years she wrote a column of Pedro Miguel events for the Star & Herald. From her present home in Romney, W. Va., she wrote:

"There was no legitimate Clubhouse in Pedro Miguel before the one from Gorgona was brought there in 1914, but we did have a social hall which was over the old messhall. This stood on the old road near what is now the junction of Gaillard Highway and Rio Grande Street.

"There our Sunday School was held, also church services whenever we could secure a minister which was once every three or four months. Once or twice a year, a Lyceum Company which was brought down from the States by the old ICC gave a program here, and here we also held our dances.

"The first Christmas program in Pedro Miguel was presented here in 1911. Miss Mildred Greene, sister of J. Wendell Greene (who retired as Treasurer of the Panama Canal Company in 1952), and I trained the children. We trimmed a small orange tree with popcorn strings and paper chains, lighted it with old-fashioned wax candles and called it our Christmas tree.

SGT. GEORGE L. CAIN, Police Commander

MISS DORA ANTILL, School Principal

"Incinerator Point—it was called that because the incinerator for the disposal of garbage was located on the point of land which jutted out into the water—was down where the SIP quarters on Frog Alley now stand, facing the lake.

"Old Man Campbell, as we used to call him—I believe his name was John—had a chicken farm there and beyond it a narrow swinging bridge crossed the Pedro Miguel River to the native huts which were grouped on the other side. There dances were held in the evenings to the music of tom-toms.

"The area where the Lodge Hall now stands and the land immediately below it was once a native village, with thatched roof huts; I remember it was a pleasant sight in the evening to see the light of their fires and to hear their voices raised in plaintive songs.

Christmas On Stilts
"On Christmas morning, dressed in grotesque costumes, they would stalk through the town on stilts. Some of the stilts were so tall the people on them could look into upstairs apartments. Later in the day, they drove or led a giant tapir up to their village where it was killed and roasted over an open fire and a great fiesta was held.

"The present Boat Club grew out of the old El Kego Club, which met in a thatched roof shed which stood near the present location of the Boat Club. The men selected this spot, isolated as it was, so that their noise would not annoy residents of the town. Here they enjoyed their keg of beer and passed the hat to defray the cost. Its fame spread until the Boat Club was organized to take care of the crowd.

"Al Meigs was one of the organizers and perhaps did more for the club than any other man. He was ably assisted by Jack Reinig, J. C. Ewing, S. B. Bubb, Adam Mallett, Adam Dorn, Harry Groschup, and many others.

MRS. JANET BIENZ, Restaurant Manager

"The old police station, which was brought from Gorgona, stood on the left of Front Street, as you walked from the railroad station to the clubhouse. In the early days, a stable for the horses stood in the rear of the building. The concrete base and the steps leading to the station are all that remain today.

"In the early days we made our own entertainment. We used to put on plays, give programs, hold dances. There was the Ladies' Aid Society and the Woman's Club. We often attended dances at Gorgona and Empire in a labor car hitched to an engine. Some engineer would volunteer to run the engine there but wouldn't guarantee to bring it back. Perhaps another one who had celebrated less would make the return trip with us."

After Construction Days
Pedro Miguel Locks were the second to be started. Concreting there began in September 1909, a week after pouring had started at Gatun. A little over three years later, Pedro Miguel Lock gates were closed and opened for the first time. THE CANAL RECORD reports that gates in the east chamber were first operated November 16, 1912; they were set in motion by the little son—unnamed—of Congressman John J. Fitzgerald, an Isthmian visitor. The first lockage, however, was not made until October 24, 1913.

As construction ended, plans were made for the future of Pedro Miguel. In June 1912, a committee appointed to choose sites for permanent townsites recommended that the operating force of the Pacific Locks should be housed in one settlement, and chose a location on the east side of Pedro Miguel Locks. As it was then planned, the town was to house 62 American and 162 alien workers and their families.

Eventually, but only after considerable heated correspondence between the landscape architect and townsite engineers, a new Pedro Miguel began to take form. The commissary, police station, and clubhouse were transferred from Gorgona; the center of population shifted from the west side of the railroad to the east. An old barracks building was converted to quarters for "lady bachelors," because, the Pedro Miguel quartermaster said, "it appears that there will always be at least seven or eight lady bachelors in Pedro Miguel."

The big house formerly occupied by W. G. Comber, Superintendent of Dredging at Paraiso, was moved to a site uphill from the clubhouse. Its most recent occupant was Truman Hoenke, Pacific Locks Superintendent. Previous occupants were the Roy Stockhams, J. C. Myricks, and John G. Claybourns.

New quarters were built along the newly-made streets; nine houses were moved to Pedro Miguel from Las Cascadas. A restaurant was opened in the building which now houses the postoffice, clubhouse luncheonette, and barber shop. Some years later a dozen cottages and four two-family houses were brought from Gaillard and Empire and rebuilt in the swampy area near the lake; it is known to all old Pedro Miguelites as Frog Alley. The present fire station was not built until 1932, and the police station, now officially the Canal Zone Prison for Women and Juveniles, in 1934.

Plans For New Town
In 1938 the Pedro Miguel Civic Council, an active group which is just now planning the third annual town fair and is, incidentally, the only council with junior members, asked for a relocation of the entire town. For two years they argued their case and finally, in April 1941, the Governor approved a plan to develop an area on the west bank of the Pedro Miguel river. This would involve replacement of 127 pre-1915 quarters.

PAUL KARST, Postmaster

LT. CHARLES F. STEVENS, Fire Station Commander

E. B. VERNER, Commisary Manager

DR. DAVID SENZER, District Physician

A few quarters were built to house people working on Special Improvement Projects for the locks, but Pearl Harbor canceled out all plans for a new town. Pedro Miguel took on the look of an armed camp. Barrage balloons flew above the locks. An anti-dive bomber net stretched across the canal between the hills. Its effectiveness was tragically proved when a low-flying U. S. plane tangled in the dangling cables and crashed to the canal bank. Baffles of corrugated metal enclosed the locks. Smokepots burned from time to time along the main streets, too often, housewives complained, on laundry days.

Air-raid shelters were built in the hills behind the town, one of them near a neglected old cemetery which, Mrs. Ewing says "did a flourishing business during construction days." Its graves and their markers are again overgrown with grass. A USO unit was set up in the basement of the Union Church; the women of Pedro Miguel took pride in learning their service friends' birthdays. There was always a birthday cake for each one and the women saw to it that he was there to enjoy it, even if it meant going to his colonel to get him a pass.

Today Pedro Miguel's future is limited. Present plans call for its discontinuance as a Canal Zone community by the end of March of next year. The old quarters there will be torn down and only a few of the permanent buildings such as the police and fire stations will remain. Eventually people may say: "There Pedro Miguel used to be."

Pedro Miguel diehards still like their town, rundown at the heels as it is; of course, they would like more modern housing. A little over a year ago 110 families—out of 184 employees—petitioned that Pedro Miguel be retained and rebuilt in its present location.

Possibly, Mrs. Ewing expresses what many of them feel:

"I have always loved the town and have gone over its streets, its homes, and all they contained again and again in my memories since I left there."

The front of the Pedro Miguel Elementary School looking south, circa 1920. *Courtesy of Hallen Collection.*

Postcard, a Street Scene in Pedro Miguel, showing the dainty cottages that families are furnished with. Photo by Underwood and Underwood, N.Y., circa 1910. *Courtesy of Isabel Wood Egan.*

Left, Relaxing in "Frog Alley," Pedro Miguel, 1940s. Right, The "Big House" in Pedro Miguel was made by combining three Construction Era houses, circa 1930-1950. *Courtesy of James W. Reece.*

Memories

Life on Incubator Row in Pedro Miguel from 1951-53

Incubator Row was a long curved street in Pedro Miguel. This was where newly-hired Locks employees from the States lived. Most of the families included babies, toddlers, and/or small children-- plus many of the wives were in a family way, including me! Hence the nickname--Incubator Row!

The quarters on Incubator Row all looked the same-- old wooden 4-family buildings with no space underneath for laundry, maids' rooms, or parking. During rainy season plans for children's outdoor activities were short and uncertain.
But dry season was different-- skies were blue and trade winds were blowing. In the afternoons neighborhood mothers spread old quilts and blankets under the large mango trees in our backyard. Soon a group gathered and while mothers chatted, babies and children played, ran and shouted.

Conversations of the mothers usually included local gossip and coping with life on Incubator Row-- there was heat, rain, noise, bugs, mildew, 25-cycle electricity, adjusting to husbands' shift work, and shopping at that big old commissary across the highway.

On the positive side-- job benefits and pay were very good, living expenses low, gasoline cheap, maids plentiful, and local foods and lottery tickets available in Panama City.

Our family left the Canal Zone after two years, but we returned eight years later and stayed for 22 additional years. With the passing of time-- we found that all the negatives we remembered about the Canal Zone had turned into positives!
- *Jackie (Ward) Forrest*

A Memorable Dip in the Pedro Miguel Pool

Pedro Miguel was Shangri-La for those of us privileged enough to have lived there. Even as very young grade school children we had liberties unheard of in today's tumultuous world. The Canal Zone was safe, and we could be gone from home all day without much concern by our parents. As a result, of course, there was mischief to be had.

One night, well after dark, a few of us boys (probably around age 12) decided to sneak into the fully fenced community swimming pool for a cooling dip. Once in the water, our excitement must have gotten out of hand and our noise attracted nearby residents, who summoned the police. There was a heck of a scramble as the police car approached, each of us running off in different directions. I ran into the pool control room and up to the top of the "chlorine tank", (as we called it) some 15 feet up. I lay down on the scaffolding making myself nearly invisible, but with a great view of the surrounding area. There, I watched the police catch every other one of my buddies. Their flashlights came my way on several occasions, but they never saw me. I watched as they put my buddies in the police car and, finally, drove off.

After a few minutes, I climbed down and headed for home. At first, I was pretty pleased with myself, but my conscience began working on me. My buddies had been captured, and who was I to go free? Fraught with guilt, I continued right on past my house and went directly to the police station where I entered . . . with my hands raised over my head.
- *Bill Hatchett*

Left, Canal Zone Police Station during the construction era.
Below, Canal Zone Police Station, circa 1939.
Courtesy of Hallen Collection.

Pedro Miguel Train Station, circa 1939.

Left, Pedro Miguel Clubhouse.

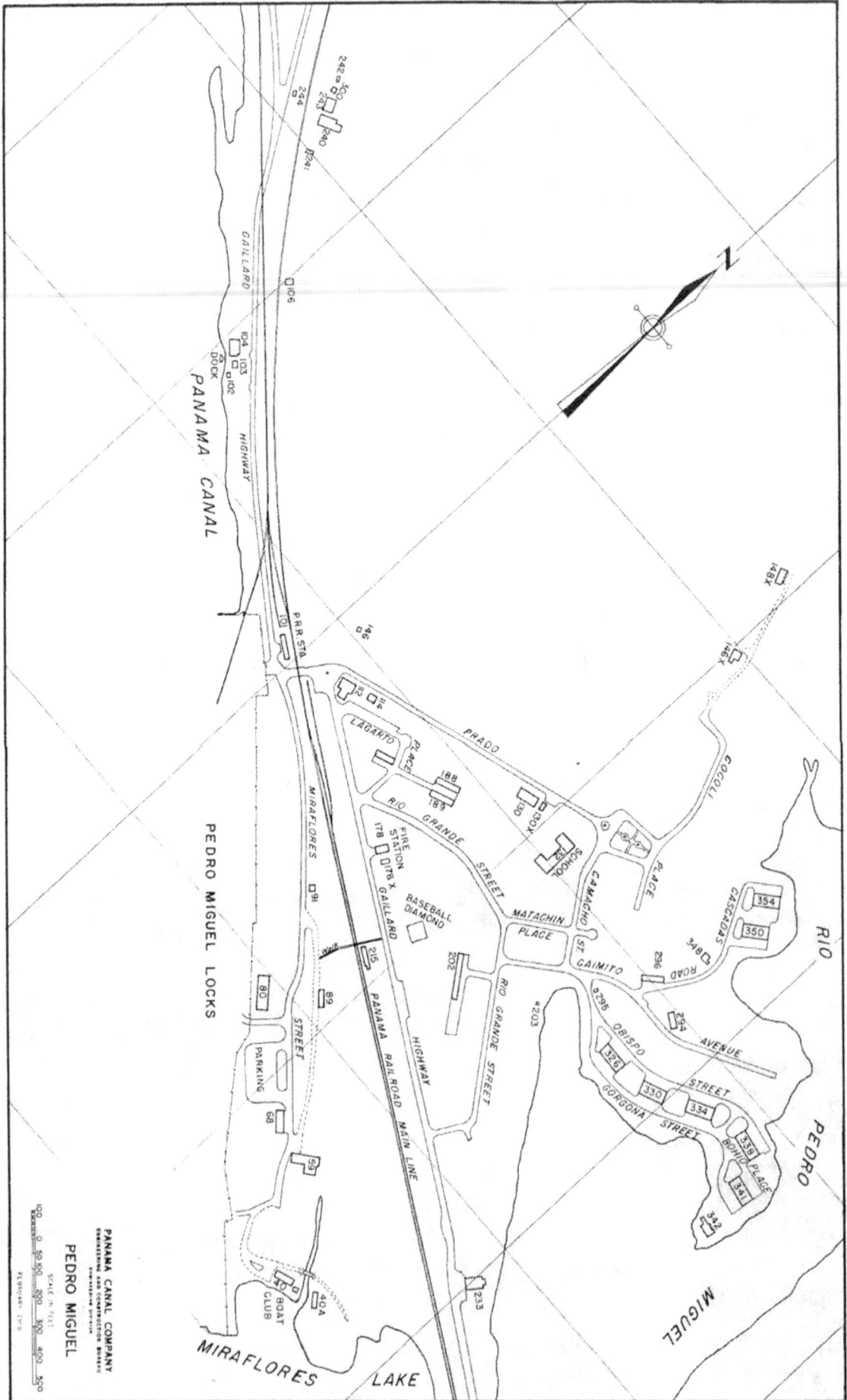

PANAMA CANAL COMPANY
ENGINEERING AND CONSTRUCTION BUREAU
ENGINEERING DIVISION

PEDRO MIGUEL

SCALE IN FEET

100 0   50 100   200   300   400   500

FEBRUARY 1978

PANAMA CANAL

GAILLARD HIGHWAY

DOCK

PEDRO MIGUEL LOCKS

PRR STA.

MIRAFLORES STREET

PANAMA STREET

PARKING

BOAT CLUB

MIRAFLORES LAKE

LAGARTO PLACE

PRADO

RIO GRANDE STREET

FIRE STATION

GAILLARD

BASEBALL DIAMOND

MATACHIN PLACE

CAMACHO ST.

GAIMITO ST.

RIO GRANDE STREET

PANAMA RAIL ROAD MAIN LINE

HIGHWAY

SCHOOL

COCOLI PLACE

CASCADAS

OBISPO STREET

GORGONA STREET

BONO PLACE

AVENUE

RIO

PEDRO

MIGUEL

126

Rainbow City

Sponsors

In Memory and Honor of the Residents of Rainbow City

Your Town - Rainbow City

The Panama Canal Review, September 2, 1955, pp. 8-10.

RAINBOW CITY is the largest of the Canal communities. Its newest section is in the foreground; some of the school buildings and part of the older section of the town appear in the upper lift. Camp Coiner is located across Randolph Road to the left, out of camera range.

A passenger on an Atlantic side bus one of the Dragnet Line, for instance never gives his destination simply as Rainbow City. That would be like telling a taxi driver in New York that you'd like to be taken to the East Side or a Boston chauffeur that you want to go to Back Bay.

Rainbow City, bisected by Randolph Road, is spread out over such an area that a bus passenger has to be pretty specific about where he wants to go: The Cantonment, Camp Coiner, the new town, or the Market.

To many Isthmians, Rainbow City is the cluster of 91 almost new masonry houses between the Folks River estuary and the cluster of two-story wooden houses which used to be known as Silver City Heights.

Actually, Rainbow City includes both these sections and the areas known until 1952 as Silver City and Camp Coiner. Few of Rainbow City's buildings are more than 30 years or so old, but its general location has been a housing area since French Canal days. Old maps show a little settlement called Guava Ridge about where the newer section of Rainbow City stands today. The main areas for the "native" housing, however, were further north.

Fox River, Camp Bierd

When the American canal builders arrived early in the twentieth century they found two main groups of houses in this area. One group was at Folks River (referred to until about 1915 as "Fox" River) and the other was on the shore of Limon Bay, overlooking Telfer's Island where quantities of French equipment had been abandoned. The first of these settlements eventually became the forerunner of Rainbow City; the latter is now known as Camp Bierd.

The Fox River settlement was apparently the older and larger of the two. It lay in the line of a sewerage layout and the houses had to be moved about 1906. At this time there are references in old files to "small portable houses put up by the French and in bad condition" and to "24 main buildings in three rows" between the railroad shops and the main line. Chief Engineer John F. Stevens had them moved onto newly filled ground so close to the border that the street on which they faced was in Panamanian territory.

The quarters at Camp Bierd-which undoubtedly got its name from W. G. Bierd, an early Superintendent of the Panama Railroad-included a few houses for families but most of the Camp Bierd buildings were barracks for dock workers; these were not the big structures which barracks are today but one-story build-ings, each housing 25 men. The messhouse at this time was an old "magazine" which had been used as storehouse for bricks. It later became a barracks in its own right and finally combined housing functions with those of a furniture warehouse and storeroom.

It is hard to determine from old records how many people lived in each of these two major settlements; the files show merely that there were 2,439 men, women, and children in "silver" quarters in the Cristobal District in 1907, and that the school, with an enrollment of 166, was the largest colored school in the Canal Zone.

Population Steady

The fact that Cristobal was already a major port and railroad center kept the population of the two settlements fairly steady, as a majority of the men living there worked in the railroad shops, on the docks, or construction. Consequently neither Camp Bierd nor Fox River underwent the decrease of population or even oblivion which occurred to many Canal towns as construction days ended.

Soon after the Canal was opened, a survey was started to determine housing needs all through the Canal Zone; The answer, as far as the Atlantic side was concerned, indicated an urgent need for more quarters, fast.

This was 1915. The population of Fox River was 932 and that of Camp Bierd, 1,818. The Camp Bierd barracks were filled beyond their capacity and 500 apartments were needed for married employees in the two sections. The Canal's Quartermaster Department began plans for a new "silver town."

Noah's Ark

To help out the acute situation the Canal took over from a private owner a huge structure of 140 rooms which was known to Atlantic siders as Noah's Ark or the Long Building. For a while it housed families, then it became bachelor quarters and finally it housed both. When the Ark was finally torn down in 1928, new houses had to be found for its 48 family and 66 bachelor occupants.

According to D'Elman Clark, who has been with the Cristobal housing office since 1920, the Long Building stood just about at the north end of the present Motor Transportation Division Corral. Later this location was used as a terminal for Isthmian Airways and for PAA's amphibian flights.

DONALD E. BRUCE, Manager of the Rainbow City Commissary, and his cocker, Pudge, are inseparable.

A fill south and east of the corral was selected as the best site for the new town to house the Folks River and Camp Bierd families. Thirty-nine 12-family houses and ten 32-room bachelor barracks were built between 1919 and 1921 on the fill which was packed hard with dirt excavated by the Army during the construction of Fort Davis.

SILVER CITY

For some time neither town nor streets were named. The settlement was referred to in Canal files as silver town and Silver Town. Eventually its residents took matters into their own hands and called it Silver City. The same thing happened with the streets, which at first were numbered or lettered. It was not long before there was an Alligator Street-now St. Kitts-and a Wall Street where the more affluent Silver Citonians lived. Wall Street is now known as Jamaica Street.

Silver City continued to grow and in 1933 it got its first suburb: Silver City Heights. The difference in altitude of the two sections is barely perceptible and the name was of local coinage. Vinesa Mundly, a Motor Transportation Division chauffeur who has worked for the Canal organization since 1914, says: "It was a pleasant name and it worried no one."

Most of the buildings in Silver City Heights, designed primarily to accommodate the families still living at Camp Bierd, were two-story 12-family quarters; a good many of them are still standing. Some are in bad condition and are about ready for Chain Singh or some other building buyer: but some, of composite construction and built three years later, sport new coats of paint and are quite presentable.

Fire Refugees

Silver City had barely adjusted to the status of a town with a suburb when it had an unexpected influx of people. On the night of April 15, 1940, flames swept through the heart of Colon, driving hundreds of families from their homes. Many of them were Canal employees. Within a few days 100 tents had gone up in rows just south of the comparatively new Silver City Heights to shelter the refugees and within a few months 36 cantonment-type quarters which still stand provided more permanent shelter. At first each house had 12 apartments. Not

CAMP BIERD, where there is now only a house or two and a cluster of bachelor barracks, was once the bustling town shown in this old picture. The men in the foreground are lined up at the mess hall.

long ago they were remodeled and today four or six families live in each of the long low buildings.

The war years brought another population increase. Most of this, however, was due to the importation of contract laborers to work on the docks and the bulk of these were housed in barracks at Camp Bierd. Today most of Camp Bierd's wartime buildings have been torn down. William Jump, of the Industrial Division, lives with his family in the remodeled Camp Bierd labor office which stands alone in a section where thousands of men once milled around; Camp Bierd bachelors live in a cluster of onetime Navy barracks which they call Vatican City.

Like other Canal towns, Silver City had its civil defense, lived with blackouts new Silver City Heights houses to shelter and under wartime restrictions. Still standing today is the town's old bomb shelter, which now houses the Camera Club and some of the activities of the schools' physical education director. Just across Randolph Road was Camp Coiner, an Army Engineer camp where, one former Engineer employee says, "men and machines nested together and the men

fought the ever-present sandflies." After the end of the war Silver City got its second suburb when Camp Coiner was transferred to the Canal. None of the Army buildings remain but some of today's houses stand on old Army foundations. Later the first "experimental housing" for local rate workers was built in Camp Coiner.

Like a Rainbow

The last expansion of the town took place about three years ago when 182 apartments, all in two-family houses, were built just about a mango-throw from the Folks River estuary. The houses were, and still are, painted in attractive pastel shades. The name of Rainbow City, which the residents themselves chose in a poll, sponsored by THE PANAMA CANAL REVIEW, was a natural. The color scheme is still being followed. The new sewage disposal plant just opened in Rainbow City, is a cheerful pale green.

The first families, among them the handful still remaining in Camp Bierd, moved into the new Rainbow City quarters in December 1951. Honor of being the new section's senior resident goes to Victor Pinta who moved into his quarters the day after Christmas that year.

Rainbow City, including all of its suburbs, is the largest of the Canal towns. Its population, according to last November's census, is 4,845, and 55 percent of them are children. There are more children living in Rainbow City than the combined populations-adult and small fry-of Margarita and New Cristobal, and the Rainbow City Junior High School and the town's elementary school are the largest of the Latin-American school system.

Today Rainbow City has just about the same facilities as any Canal town, although its residents receive their mail at Cristobal. Its housewives have a choice of shopping at the big concrete commissary, built in 1930, near the Mount Hope stadium, at the Camp Bierd Commissary which was opened in 1942 when Camp Bierd was not the deserted village it is today, or at Rainbow City's own little market where a dozen or so vendors sell bananas and plantains and such foodstuffs.

Its people have a wide choice of churches. There are seven in Rainbow City and a Church of God and an Episcopal rectory are being built.

SCHOOL PRINCIPALS, William Wilkie, left, of Rainbow City Elementary School, and Owen B. Shirley of the Rainbow City Junior and Senior High Schools, are responsible for over 1,650 students.

For recreation they have, in Rainbow City proper, a handsome swimming pool with a luncheonette handily nearby and at Camp Bierd Service Center, one of the largest movie halls in the Canal Zone.

There is a Lodge Hall where the Foresters and other fraternal organizations meet, in one of the few remaining Silver City buildings, and the International Boy Scouts have a clubhouse of their own. Offices of the Cristobal Chapter of the C10 local and the Cristobal Federal Credit Union are near the Camp Bierd Service Center.

The people of Rainbow City are avid ball fans; they have both big and little

CUTHBERT C. ROWE, Manager of Camp Bierd Service Center and the Rainbow City Luncheonette.

JOHN C. WALLACE
Manager, Camp Bierd Commissary

leagues. Their team, the Dark Millionaires, is this year's champion for the Softball League and the Silver City Heights youngsters won the pennant in Rainbow City's own Little League. Joscelyn Evering is president of the Rainbow City Major Softball League; the president of the Little League is Earle G. Moore.

Cricket is, of course, another favorite sport. For the second time this year the Surrey Cricket Club of Rainbow City won the championship of the cricket league, winning 32 out of 35 games.

The Summer Recreation program which keeps the young'uns occupied during their school vacation, is carried out under the leadership of volunteer workers. This year's program was one of the most successful of any to date.

Besides its fraternal bodies and its ball teams, Rainbow City has a number of other active organizations. There is the Civic Council, whose current president is J. J. Joseph. It meets regularly to discuss the town's problems and its representatives attend the monthly "shirt-sleeve conferences" with the Governor.

One of Rainbow City's unique organizations is the Women's Industrial Club, headed by Mrs. Ethlin Belgrave. There women who are experts in some line like cake-decorating, embroidery or dressmaking, give lessons, without charge, in their specialties. Another women's group is the Atlantic Army Mothers, of which Mrs. James Chambers is president. As the name implies, they are a home front for the Rainbow City boys in the service.

Rainbow City's people are proud of their record of men in the armed forces. A larger number of young men from Rainbow City served in Korea and the first Isthmian to die there was Gilbert D. Francis, a Rainbow City boy.

The people of Rainbow City are also great civic boosters. Listen to the Civic Council President:

"We are proud of our improved and modern classrooms, our swimming pool and its luncheonette with a handy mail box nearby. We hope someday to have a large library in a building of its own but in the meantime we are very pleased with the branch library in the high school. We'd like someday to have a movie theater in our community proper. We are proud of our gymnasium and our modern commissary. We like the idea that we named the town and, well, we just like Rainbow City."

Residents Of Silver City Will Vote On New Offical Name Of Their Town

The Panama Canal Review, April 4, 1952, pp. 3, 7.

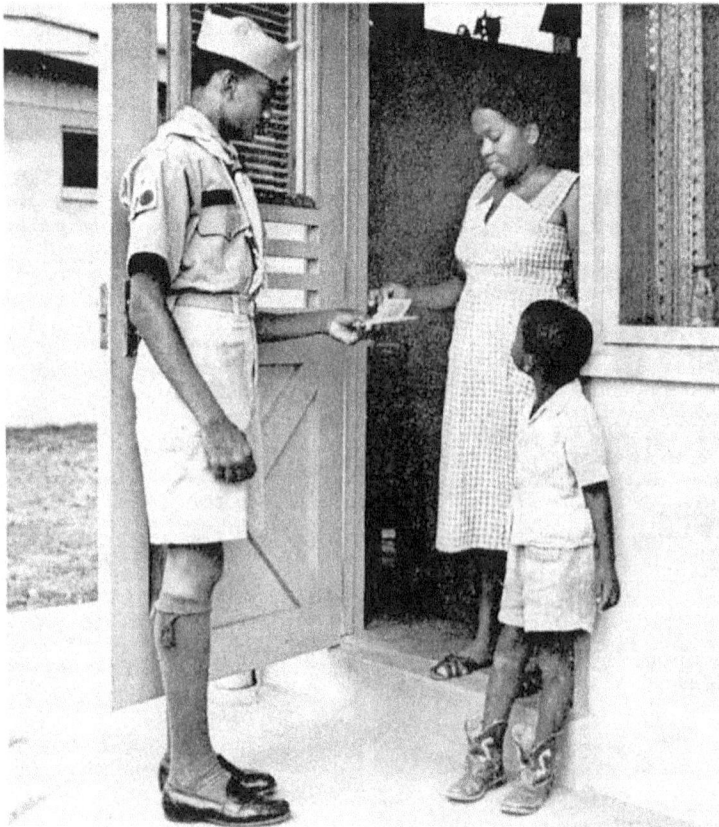

For the first time in the history of the Canal Zone the residents of a community are to be given an opportunity to vote their preference for the name of their town.

Governor Newcomer has authorized *THE PANAMA CANAL REVIEW* to conduct a poll of the residents of Silver City (including Camp Coiner) to determine if they wish to change the name of the town. The poll will be conducted this month by *THE REVIEW* in collaboration with the International Boy Scouts by a house-to-house canvass.

Six names have been proposed: Silver City, Rainbow City, Folks City, Manzanillo, Granada and Mindi. Silver City has been one of the fastest-growing towns and the nickname "Rainbow City" has been in general use by residents of the new section of Silver City as being most descriptive of the varied colors of the new houses. The other names have been chosen for the following reasons: Mindi and Folks City for the two small rivers in that area; Manzanillo, for the island on which Cristobal-Colon was built; and Granada which was the former name of Colombia.

During the time the town was being constructed, several different designations

130

were used. Among these were: The Folks River end of Manzanillo Island; silver town at Mount Hope; new silver townsite at Big Tree; and Cristobal Silver Townsite.

After the area was occupied, it was variously called Silver town, Silver Town, and Silver city. The name "city" in Silver City was not capitalized until July 1921 when used in some official correspondence. Thereafter, however, all official correspondence referred to the town as Silver City, although the files do not indicate it was formally named as were other towns of the Canal Zone, such as Frijoles, La Boca, and Balboa.

It is presently planned to distribute ballots April 16 to all employees who have assignment to quarters in the area. The ballots will be distributed and collected by the five International Boy Scout Troops of Silver City.

Governor Officially Names Rainbow City
After Overwhelming Vote of Residents

The Panama Canal Review, May 2, 1952, p. 4.

By an overwhelming majority, residents chose the most colorful of six names offered for voting. The election was very popular. Of the total of approximately 1, 280 ballots distributed, all but 30 were returned. The change in names was made effective May 1.

Modern
Civic
Center
Planned

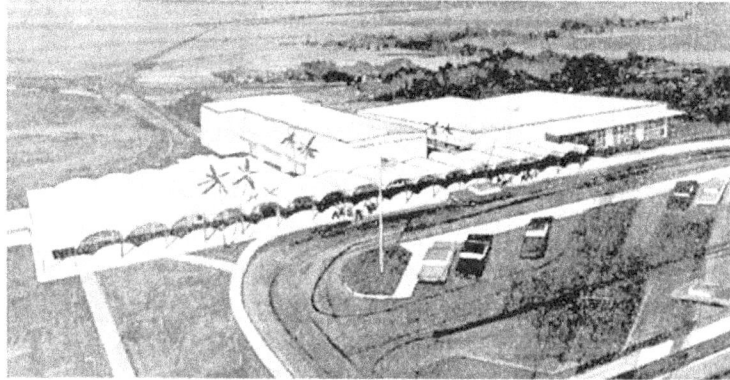

Rendering of the Rainbow City Auditorium with plans for a service center, health center, post office and aother facilities to be built adjacent to it. It has the capacity to seat 300, available for stage shows, school productions, musicals, and public meetings as well as regular motion pictures. *The Panama Canal Review,* August 4, 1961.

New Rainbow City Auditorium-Theater was opened in facility improvement program. *The Panama Canal Review,* January 6, 1961.

Rainbow City pool. *From Recreation in the Panama Canal.*

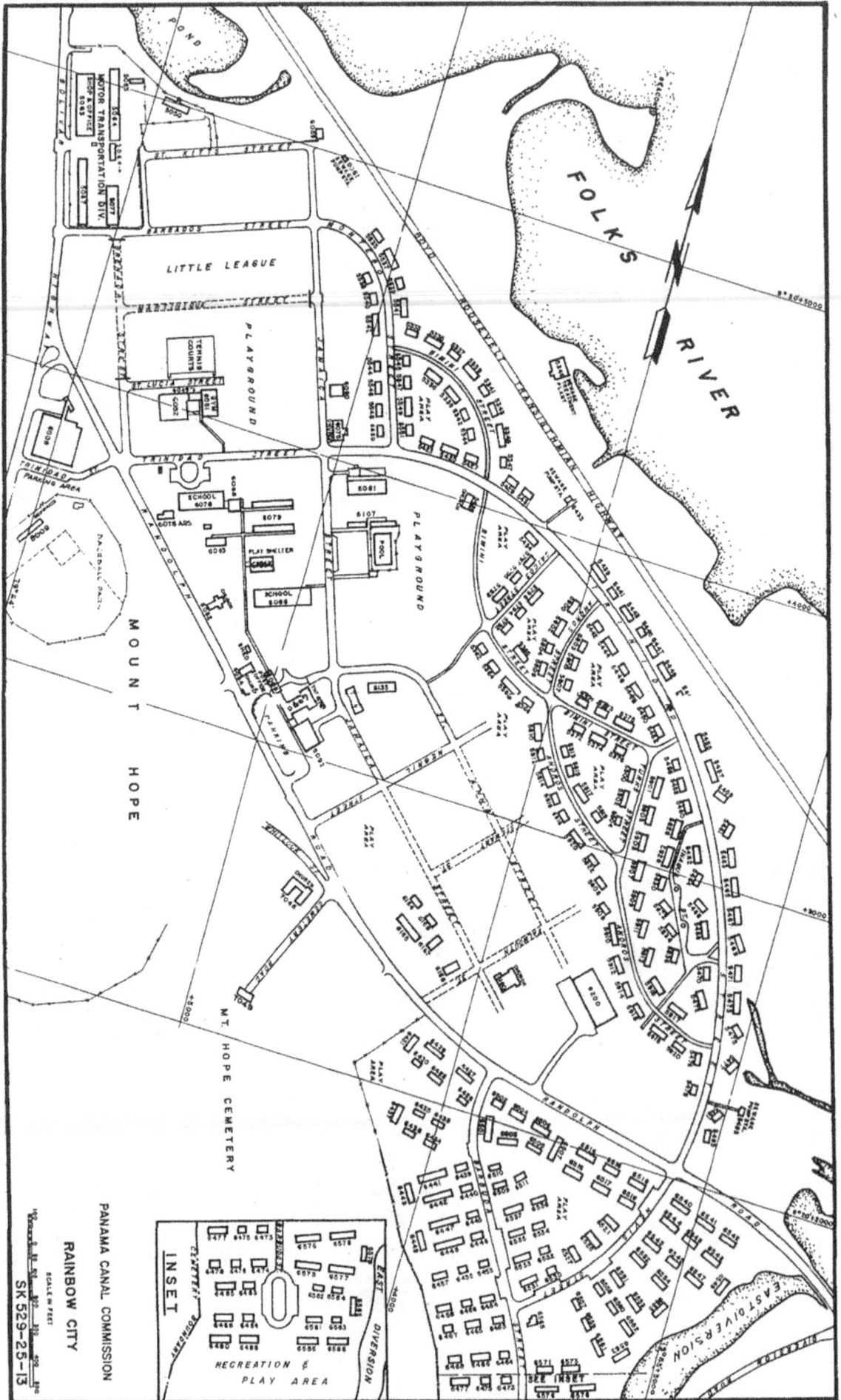

FOLKS RIVER

LITTLE LEAGUE

PLAYGROUND

TENNIS COURTS

PLAYGROUND

SCHOOL 6078

SCHOOL 6086

PLAY SHELTER

MOUNT HOPE

BASEBALL PARK

TRINIDAD PARKING AREA

MT. HOPE CEMETERY

EAST DIVERSION

EAST DIVERSION

INSET

RECREATION & PLAY AREA

SEE INSET

Red Tank

Sponsors

In Memory and Honor of the Residents of Red Tank

Ghost Town - Red Tank

The Panama Canal Review, December 4, 1953, p. 16.

SENTINAL for a deserted village, this big old Cuipo tree has been a landmark at Red Tank for many years. It stands at the end of a causeway over an arm of Miraflores Lake.

There are no more games of "Brown Girl in the Ring" on the sidewalks of Red Tank. The heated domino and draughts tournaments are no longer going on under the houses which have faced Gaillard Highway for 35 years. The market women from Chiva Chiva have stopped selling their plantain and yuca and yam at their little makeshift stands along the street.

For there are no more men or women or children living in Red Tank. The local-rate town, whose population swelled to over 2,200 in the decade between 1931 and 1941 is now deserted.

The last residents, Mr. and Mrs. Charles Moseley, moved November 12 to Paraiso where many of their former neighbors had preceded them. A Panama Canal employee for 40 years, he had been the Salvation Army's Red Tank representative for the past three years.

Birds and a Dog

For two weeks, the Moseleys were the only people living in Red Tank. Rover, whose bark is worse than his bite, and the birds which came each morning to the Moseleys' back porch to be fed, were the only other living creatures in the town.

The last Canal families had moved October 29 to La Boca. The Red Tank school which last year had 371 pupils had only two left when it closed its doors for the last time October 30. The Red Tank Commissary sold its last goods October 31.

Today Red Tank is a ghost town. Long pieces of wood have been nailed across the doors of the vacated houses which will soon be demolished. The remaining stock has been removed from the commissary. Along the back streets of the town the grass is beginning to grow high.

Pedro Miguel Tank

Red Tank's beginnings are hazy. A 1904 timetable for the Panama Railroad shows a stop called Pedro Miguel Tank, five tenths of a mile south of Pedro Miguel proper. The same timetable, which lists the tank as a stop for all trains, indicates that it had a siding for 24 cars.

Oldtimers, like William Jump, recall that there was a big water tank, painted with red lead, on a hill behind what later became the town. From this undoubtedly came the name of Red Tank which is mentioned in a 1908 file in a letter recommending the demolition of three old "buildings at Red Tank . . . they are all in very bad shape."

The name of Red Tank does not appear again in official files until November, 1915, six months after a three-man committee was appointed to investigate and report on the number of quarters which would be needed for local-rate employees near Pedro Miguel and Miraflores. The committee recommended the construction of 80 apartments, to cost $56,000, and to "be located in the vicinity of the tunnel dump."

Census For 1916: 242

The first Red Tank quarters were completed that same year. The first occupants were 42 families and 42 bachelors, all the men employees at Pedro Miguel Locks. The first census report for Red Tank showed 242 residents in June, 1916.

In 1917 more quarters were built and 83 families and 40 bachelors were moved into Red Tank from Rio Grande. Later that same year Wards 7, 8, 9, and 10 from Ancon-now Gorgas--Hospital were re-erected at Red Tank as building 536. This huge structure, housing 48 families, immediately and unofficially was christened the Titanic. The smaller building next door, which had also been an old Ancon Hospital ward, quite logically was known as the Iceberg. Both buildings had been built at the hospital in 1907. They were torn down in 1951.

By 1919 Red Tank's population had grown to 1,302 and six years later had increased to 1,672. In August, 1927, four buildings were brought to Red Tank from Culebra and re-erected as 10-family quarters. Later that year three old Army barracks from Camp Gaillard on the west side of the Canal, were rebuilt into two 12-family quarters. These last were to house local-rate employees who were still living on the west side of the Canal in the Gaillard and Empire districts.

War Boom

Like all Canal towns, Red Tank mushroomed during the hectic days just before and in the early part of World War II. Barracks were put up for local-rate bachelors, the clubhouse enlarged.

The clubhouse, which had been built in 1919, was completely destroyed by a fire on February 23,1945. The fire was caused by a break in the fuel supply line of a pressing machine in a tailor shop in the clubhouse basement. One woman, an employee in the shop, was badly burned. The heat from the burning one-story building was so intense that a wooden retaining wall across the street 'and along the railroad tracks was set afire and grass began to blaze beside the tracks. After the fire, clubhouse facilities were provided in an old school building.

A dispensary, Red Tank's first, was opened in June,1946, but closed three years later when the town's population had begun to drop.

Deserted today, Red Tank had had 1,075 inhabitants when this year's police census was taken in June. Over half of these were, children.

It was the children the Moseleys missed most during their two weeks as Red Tank's only residents.

"We were so used to the patter of children's feet," Mrs. Moseley said, "that after we were alone I thought time after time I heard children running, although I knew truly that there were no children there. I would go to the window and look. Of course there weren't children, or anyone else."

CHARLES MOSLEY and his wife missed the sound of children at play. They were the last residents of Red Tank. When they moved to Paraiso, the town was abandoned to the wild things which still come down from the hills.

Buildings Erected for Laborers and "Silver" Employees

The Canal Record, December 25, 1907, p. 134.

by P. O. Wright., Jr., Architect.

QUARTERS FOR UNMARRIED EMPLOYEES

As living quarters for the unmarried employees, five types of barracks have been built. It has been found by experience that two of this number answer all of the purposes required and construction is therefore now confined to these two types. They are designated as "Standard Laborers' Barracks Nos. 1 and 2," both having practically the same arrangement but differing in size. The standard barracks No. 2 is a one story building, consisting of a dormitory 35 feet wide by 50 feet long, with verandas 8 1/2 feet wide at each end. The upper portions of the walls of the dormitory are left open for a distance of 3 feet below the ceiling line for ventilation, and the roof is provided with a large ventilator. This opening is protected against rain by the projection of the roof, and is screened like the verandas with fine copper mosquito screening. The windows are provided with screens incapable of being opened and with batten shutters opening inward. The screening of verandas as well as of the doors is protected against careless or willful injury by slats placed for a sufficient height to accomplish this. There are never more than two entrances to the building, one at each of the screened porches, thus reducing to a minimum the chances of the introduction of mosquitoes. These barracks are provided with cursor standee berths and are intended to accommodate 40 to 80 men, allowing an average of 560 cubic feet of air per man. In case it should become necessary to increase the number of men, there is always a sufficient supply of fresh air on account of the circulation produced by the openings at the top of the walls, the end verandas and the ventilator, so that in any case, no matter what the crowding might be, the air could not become foul.

In order that the employees may have clean and dry clothes, the present Commission has been erecting buildings in the encampments where the laborers can do their own washing and have their clothes dried by steam. This arrangement is not only a great convenience which is appreciated by the men, but a precaution which greatly minimizes the sickness.

KITCHENS AND MESS HOUSES, ETC.

Food for the laborers is provided and dispensed in two types of buildings, a standard kitchen for the West Indians and a mess hall and kitchen for the European or Spanish laborers. The standard kitchen is a building 18 feet wide by 36 feet long having a concrete floor which can be washed out daily. Walls of corrugated, galvanized iron are provided on three sides while the

These 48-unit family quarters, called the "Titanic," were built for Silver Roll employees.

fourth is left entirely open and has a wide shelf or counter running the entire length of the building. Here the laborers lineup and receive their food. Above the top of the three walls of corrugated iron there is an air space three feet high, the opening being protected from rain by the projection of the roof. With this arrangement the kitchens are always cool and can be kept in an absolutely sanitary condition. The Europeans' mess hall provides, in addition to the kitchen, a large dining room where the laborers are served with food.

Bath houses are arranged so as to contain either ten or twenty cemented apartments for shower baths, and are located at convenient points in the encampment, as are also the range closets.

QUARTERS FOR MARRIED EMPLOYEES

The married quarters for laborers or "silver" employees are one-story buildings, long and narrow, with verandas in front and rear running the entire length of the building. The apartments usually consist of two rooms, and the veranda is utilized for cooking purposes, being provided, opposite each apartment, with a locker and an arrangement to hold a charcoal brazier. The toilets, as in the case of the bachelor quarters, are in separate buildings.

Quarters For Biggest, Smallest Families Planned
in Current Building Program

The Panama Canal Review, May 2, 1952.

To accommodate the larger than average families, 10 percent of U.S.-rate quarters will have four bedrooms per unit, and 15 percent of the housing to be built in local-rate communities will be of the four-bedroom type.

Housing for large families in the Canal Zone is not a new problem. Even before the present housing program was started several projects were considered to provide such facilities.

A large living room was created on the first floor by removing partitions, and the two ground floor kitchens were combined into one. On the second floor, kitchen equipment was removed. Partition walls were soundproofed.

The experiment was not considered a success and has not been repeated. The center apartment, made from the former four central apartments, was less desirable than those on the two ends. The building was of the temporary construction type and expensive to maintain.

More recently, plans were studied for converting some of the type 215 (one bedroom, four-family) houses in Cocoli. A duplex house with three bedrooms in each unit was to have been made from a former four-apartment house. This plan was abandoned as uneconomical.

Now, however, with the number of four bedroom quarters in the new housing program, it is believed that adequate living space can be provided for those families who require large quarters.

Floor Plans for Large Family Quarters

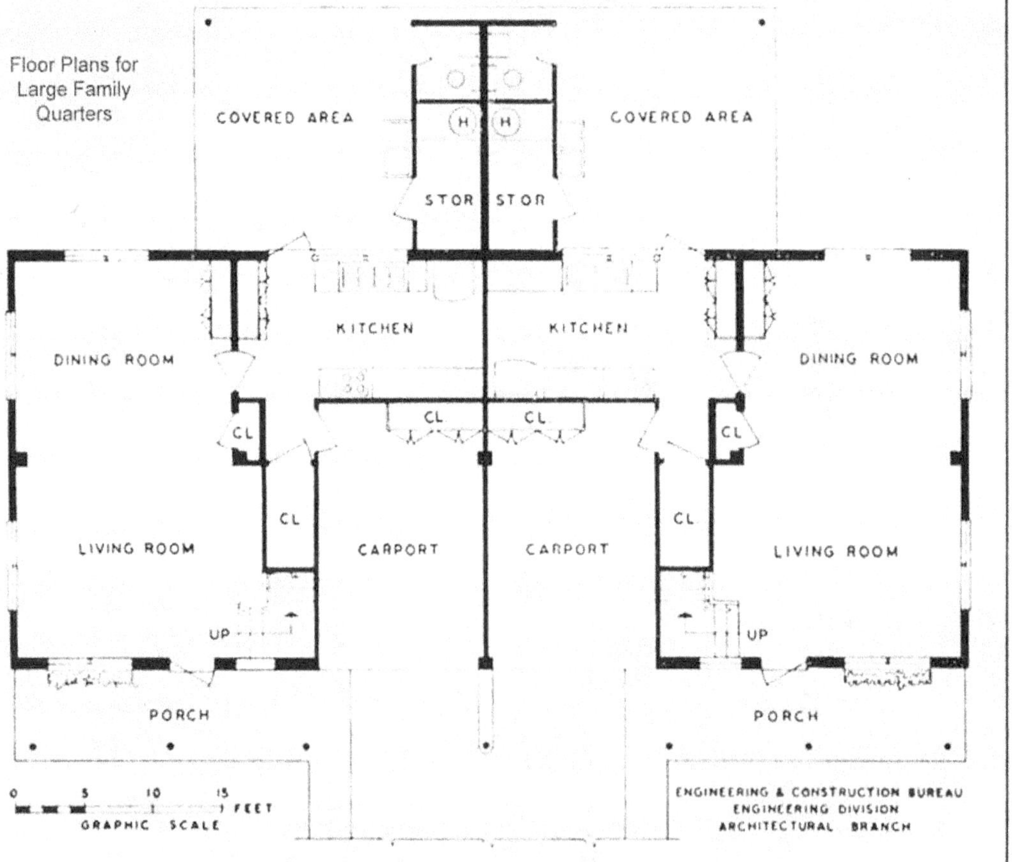

FIRST FLOOR PLAN

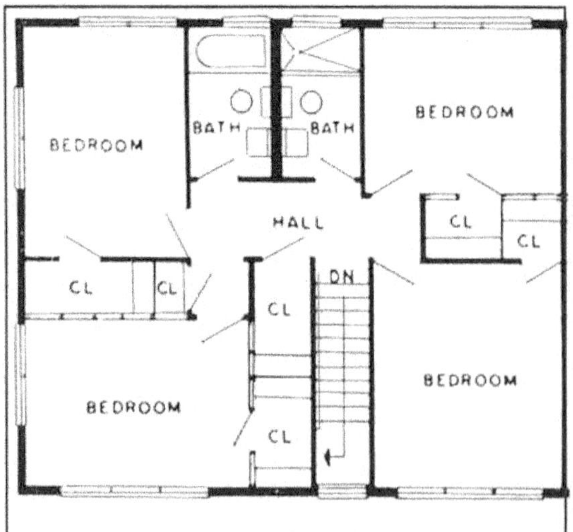

SECOND FLOOR PLAN

New Rule on Quarters

The Panama Canal Review, March 7, 1952, p. 7.

Residents of Pedro Miguel will be permitted to apply for and receive assignments to quarters in the Balboa-Ancon-Diablo Heights area after the first of next month.

The change in the rules for quarters' assignments was made in conformity with the Housing Division's overall policy to remove undesirable or inequitable restrictions on employee residence in all possible instances. Pedro Miguel residents have requested the change in the housing assignment rule on numerous occasions and their requests have been backed by the Chief of the Locks Division whose employees are mostly assigned there.

The change will become effective after the present "freeze" on assignments in the terminal towns of the Pacific side is lifted. The restrictions on quarters' assignments were placed in effect after an agreement was reached for the transfer of Cocoli to the Navy. They were to have been removed at the end of February but were extended for one month.

Military Townsites

Fortifying the Canal

During the construction phase, the question of whether or not the Panama Canal should be fortified was debated by the American public, Congress, and the world at large. The Treaty to Facilitate the Construction of a Ship Canal (more commonly referred to as the Hay-Pauncefote Treaty) established that "the [Panama] canal shall never be blockaded, nor shall any right of war be exercised nor any act of hostility be committed within it." Ships of all nations were to have equal access to the canal, during both war and peacetime, and in order to guarantee equal access, the United States was "at liberty to maintain such military police along the canal as may be necessary to protect it against lawlessness and disorder."

Albrook Air Force Base

I remember that I used to play behind the General's house where there was a golf course. We used to pretend like we were riding horses. We played in the sand traps on the golf course; we made all kinds of sand castles. We used to have this place called Purple Mountain because the clay was purple. We made a fort on this hill where we used to go everyday after school and play the rest of the day.

We used to make homemade kites from a bamboo tree and construction paper. In my yard there was a big mango tree, and we made a tree house in it with a big tire swing. There was a big hill up the street from where I lived. We used to take cardboard boxes and slide down the hill in them.

I remember some of the games we used to play like Patty Cake and a bunch of other clapping games. We played Hide and Seek at night. We also played Red Light, Green Light; Red Rover and One Potato.

My grandparents lived in Los Rios so we used to collect Christmas trees from Albrook and take them over there. We didn't have a Christmas tree bonfire in Albrook.

We used to go skimboarding in the rainy season in this open field that always used to be full of water after it rained. We also went gutter sliding in the rainy season and had mud fights.
- *Paige Heddaeus in 1978 for a folklore class taught by Mary Knapp at Balboa High School.*

BOQ, Albrook Air Force Base, C.Z.

Curundu
Curundu Heights
Sponsors

Bob Russell

Curundu's School of Tomorrow

The Panama Canal Review, March 1964, p. 6.

The town of Curundu, then known as Skunk Hollow, was established in 1919 as one of the largest military installations in Panama. Curundu had family housing, communications sites, community administrative buildings, Curundu Elementary School and Curundu Junior High School. In 1965 a multimillion dollar school plant for Curundu Junior High School was built with a geodesic dome, an architectural form originated by the noted U.S. architect, R. Buckminster Fuller. The dome housed the cafetorium designed as a combination cafeteria and auditorium. The school accommodated 2200 US-citizen students on the Pacific side in grades 7, 8 and 9.

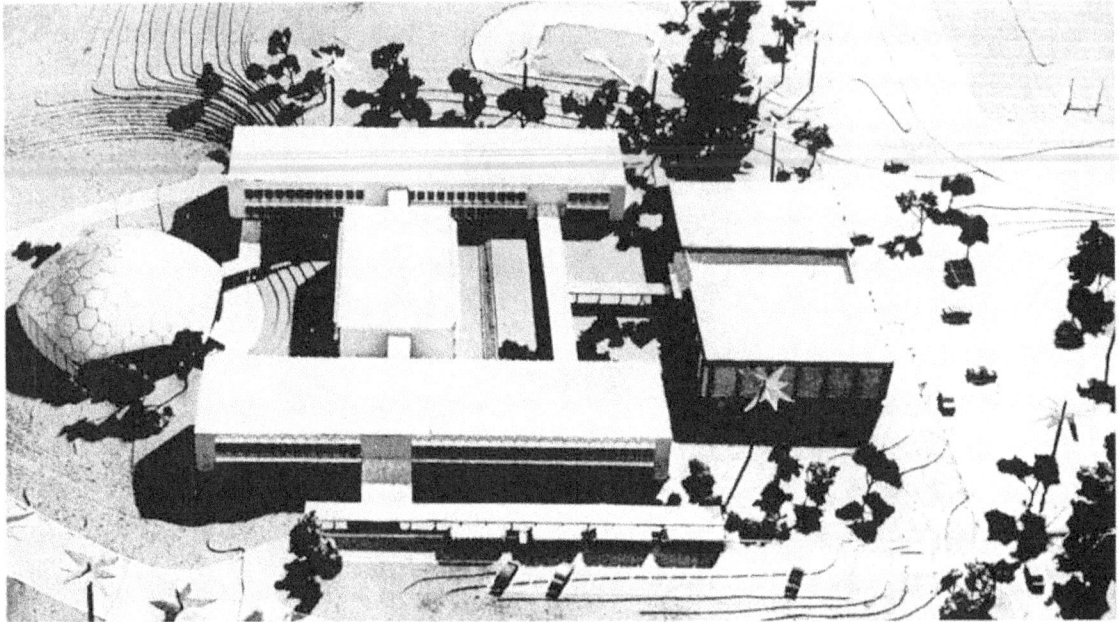

A model of the Curundu School with geodesic dome at the left.

What's in a Name

The Panama Canal Review, Spring 1972, pp. 12-13.

As were Rainbow City and Los Rios, Curundu was named by popular vote. The town was known as Skunk Hollow; but in the 1940's some residents wanted to change it to Jungle Glen, while others were for keeping the name of Skunk Hollow.

An editorial in The Star & Herald of March 18, 1943, was in favor of retaining the name stating: "Friends of tradition and Skunk Hollow need to arouse themselves if they want to save the name. They deserve encouragement. This world tends to become a dreary and orthodox place. Whatever piquancy and humor is inherent in the name of Skunk Hollow should be

preserved for the coming generations. They, to whom the old place has the associations of home and friends, cling to the old name. They might agree that a rose by any other name would smell as sweet but not Skunk Hollow."

A letter to the Panama American urged compromise by agreeing to such a name as Jungle Hollow or Skunk Glen, preferring the latter. The problem was solved by ballot and a headline announced the result "Skunk Glenners Vote Overwhelmingly for Name Curundu." Curundu was the name of the little river nearby. It is a historic name which has been spelled a variety of ways, but the exact meaning is not known.

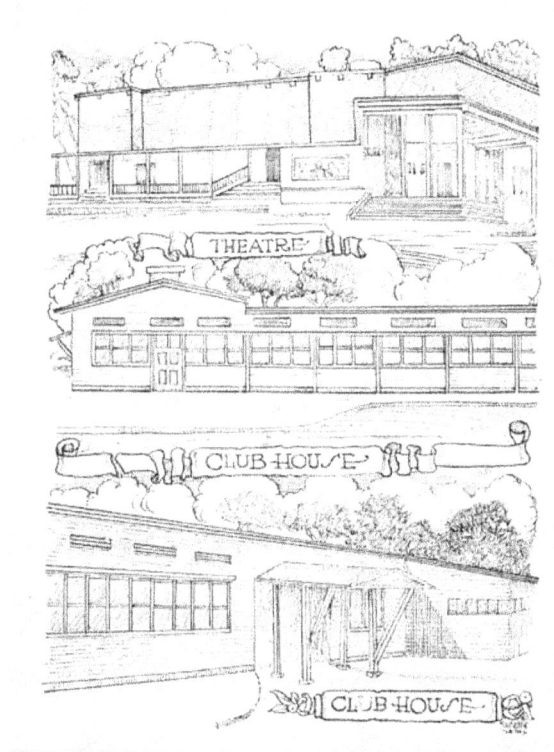

AUTUMN, 1939.   Machete gang starts clearing site for Curundu Heights.

FEBRUARY, 1940. Footings poured for quarters.

MAY, 1940.   First residents move into frame quarters at Curundu Heights. Among
                 them are Raymond V. Parker, now heading the Electrical Unit, Engi-
                 neering Branch, Panama Engineer Division, Tivoli Office, and George
                 Foote, of Supply Branch, Corozal. Marguerite Riley, Chief of the
                 Fiscal Sub-Unit, and Viola Riley, of the same office, follow shortly.
                 Early residents refer to area as "Buzzards Roost," because during
                 first weeks they lived without gas, electricity, ice.

AUTUMN, 1940.   Acute shortage of quarters for employees of War Department. Plans
                 are made to relieve this and provide for incoming employees of the
                 Panama Air Depot, then already under construction. A "Low-Cost
                 Housing" project is authorized and site selected in uncut jungle on
                 Albrook Field for the community now called "Curundu." Construction
                 begins in November, with the building of the main highway.

FEBRUARY, 1941. Construction of "Low Cost" quarters starts and moves briskly.

SPRING, 1941.   Bachelor quarters occupied though doors are missing, there are no
                 screens in windows, and nearest running water is a construction
                 hydrant two blocks down the road. Insects, iguana, snakes, coati,
                 monkeys, wild pigs and other creatures still sharing the area cause
                 inhabitants to refer to it by comic-strip names. Among the more
                 popular are "Punkin' Center" and "Skunk Hollow," the latter a neigh-
                 borhood well known to friends of "Lil' Abner."

JULY 2, 1941.   Leonard T. Collins, his wife, and eight children become first occu-
                 pants of family quarters. He was then working with construction
                 crew, is now Carpenter Foreman with the Manufacturing Branch, Main-
                 tenance Section, Panama Engineer Division. Louise Vargas, still in
                 the Housing Manager's Office, came the next day, and Neuman Dudrow,
                 Plumbing Foreman, followed shortly.

AUGUST, 1941.   Rene Espejo, Chilean artist, now with Manufacturing Branch, Corozal,
                 paints a portrait of Odie Skunk as a decoration for the front of the
                 bachelor barracks. Every taxi-driver in town now knows the road to
                 "Skunk Hollow."

DEC. 6, 1941.   Six-sevenths of quarters in present Curundu are substantially com-
                 plete except for grading, roads, walks, drainage, etcetera.

MILESTONES

DEC. 7, 1941.   Japanese attack Pearl Harbor. Machinery, men and materials are diverted from present Curundu to more urgent projects.

S. J. Sperondi, Chief, Purchasing and Contracting Unit, Panama Air Depot, and R. L. Hamblen, Principal Storekeeper, become first PAD residents of Curundu.

JANUARY, 1943.   Urgent projects have progressed to point where basic plans for community can proceed. Drainage problems are analyzed in greater detail than at first contemplated.

FEBRUARY, 1943.   Some 200 women employees of the Panama Engineer Division arrive from the States and are quartered in the Northern section of present Curundu. Name of area is changed from "Skunk Hollow" to "Jungle Glen" in official mail.

MARCH, 1943.   Curundu Heights, Diablo Terrace and area surrounding present community center designated as civilian areas by order of Commanding General.

SPRING, 1943.   Controversy breaks out over whether name of area is "Skunk Hollow" or "Jungle Glen." Colonel George Mayo, Division Engineer, appoints committee to consider suggestions, recommend a permanent name to the Commanding General.

Community center buildings are designed by architects and engineers of the Panama Engineer Division, who carefully coordinate design with needs of using agencies. Theatre is started April 14th, Clubhouse May 4th, Post Office May 11th and Commissary June 4th.

Name "Curundu" overwhelmingly approved at mass meeting; it is submitted by committee and adopted by Commanding General. Street names are chosen and signs put up at corners.

JULY, 1943.   Curundu Civic Council chosen by formal vote of residents in Curundu, Curundu Heights and Diablo Terrace.

AUG. 31, 1943.   Theatre, Commissary, Post Office and Clubhouse - including a ballroom, bowling alleys, game room, cafeteria-restaurant, soda fountain, newsstand, barber shop and beauty parlor - are ready for operation following the Grand Opening tonight. Parking facilities and roads are ample, drainage is provided for, sidewalks are laid on Main Street, basic landscaping is finished and the grounds made ready for residents to complete the landscaping of the whole area according to master plan.

 Memories

Memories of a Curundu Piano Tuner

Around 1960, when we lived in Curundu, we got a second-hand piano that needed tuning. Balboa High drama teacher Don Musselman recommended the piano tuner Federico Garcia Vallejo. He had fled to Panama after the Spanish Civil War and eked out a living tuning pianos, so he was always on the lookout for more business.

He was an expert on the bandurria, a kind of mandolin, and once gave a concert at the JWB [Jewish Welfare board]. And he founded the "estudiantina" musical group at the University of Panama that now bears his name.

He had been a friend of the Spanish poet Federico Garcia Lorca and had played for a production of one of his plays. When Don Musselman directed a bilingual production of Garcia Lorca's "Blood Wedding" that both Leonor and I were in, he helped with the background music.

I think Garcia Vallejo kept track of every piano in the Canal Zone. One day he came to re-tune ours. New neighbors, the Mullers, had just arrived from the States. When Garcia Vallejo was walking away from the building their daughter Lessie started to play her piano. He whipped around to see where the sound was coming from and exclaimed "There's no piano in those quarters!"

A glimpse of another man's life far from home; a link to the Spanish Civil War, a sense of a whole different world just across the border in Panama, all from our louvered wooden windows on 5th Street, Curundu.

143

Teaching Phys Ed at Curundu Junior High

"Come on down," (or words to that effect) said the telegram from the Panama Canal Company that bright and sunny morning in 1963—we are offering you a job: What an opportunity for a young physical education teacher barely out of college!

Adventure, romance, travel and an increase in salary (which all came true). In my state of euphoria I hardly remember the trip from New Orleans, where the Company owned ship picked up the teachers and other C.Z. employees.

All the single teachers were assigned the same type housing—1930's 3-story word shot gun apartments, which meant we had to walk through the bedroom to get to the kitchen. We took it with grace and good spirits.

My school was Curundu Junior High, a brand new school which opened that year. The facilities were wonderful—a gym for the girls and a separate one for the boys. A swimming pool and all kinds of equipment were provided. No discipline problems as the parents were held responsible for their children's actions. (What a concept!)

I met my future husband at the Balboa Yacht Club. He had a 36-foot wood ketch and a dream of sailing Shearwater around the world.

With a sense of adventure and trepidation, we left the Canal Zone in February 1968 to catch the trade winds south—south to Tahiti!

-Silver Crossman

Discovering Panama

I came to the Panama Canal Zone via the Penn State Hilton as one of 14 young teachers in the fall of 1978—all 21 years of age. I was assigned to teach 7th grade art at Curundu Junior High, under the guidance of Bill Koons, the department chair. My two friends, Joy and Becky, taught art as well—Joy was at Curundu with Henry Barker and Becky was at the high school. Henry became our primary tour guide and chauffer each morning as we piled into the back seat of his very compact Pacer -- shared with his wife and children.

Our first organized trip was to see the locks at Miraflores; it cost all of one dollar to take the Panama Canal Tour that September. Early in the month, Jim Freeman, Assistant Director of Operations, took several of us in his black and dusty 1973 Russian jeep climbing up the side of a mountain like a billy goat to a hotel where we played dominoes and ate plantains until the torrential rains allowed us to journey onward. We traveled to Las Estrellas to meet Edward Pike, a generous and genuine Panamanian gentleman with snow white hair whose lovely house windows looked down upon what seemed like all of Panama. He had erected a 12-foot concrete cross upon a distant high mountain—with a view that attracted many a visitor—and that day I saw my first double-rainbow.

The following week we went to the San Blas in a tiny 9-passenger 120 mph plane that flew only 1200 feet above sea level. Smoother than my mother's driving, our pilot, Carlos, got us safely to one of the 51 Kuna-inhabited islands. Staying in a bohio (thatch hut) and sipping a very potent Ron Cortez rum and coke while enjoying the resident parrots was the most difficult activity in which we engaged. Next stop was the volcanic beach at San Carlos where I had a close encounter with a rather large and ugly iguana in my bed. It is a time in my life that I will always cherish with a place and people that were like no other.

- Dr. Rachel A. Schipper, Associate Dean, University of Florida Libraries

Farfan Naval Radio Station

Sponsor

Charles W. "Chuck" Hummer

Implementation of the 1977 Panama Canal Treaty: An Overview, p. 23.

The Farfan Housing Community, established in 1947 and covering 860 acres, includes 78 family units, 5 steel radio communications towers, and a two-story bombproof operations building.

Dahlia Lane. *From Implementation of the 1977 Panama Canal Treaty.*

Memories

Hanging Out in Farfan

We used to hang out at the basketball courts by the movie theater in Farfan. We'd swim at the pool during the summer. We used to grab a piece of cardboard and slide down the hill. Some nights you could hear the AP's patrolling with their dogs. We had a corner house with a big side yard and played football there.

- Spencer Plantier

Farfan Movie Theater

Far-Fan was navy radio instalation, on a porton of Ft. Kobbe. As kids, most of us like to stay there, instead of going to swim/ watch movies at Fr. Kobbe.Our theater was small, and a bit old, but it was ours and, by golly, we were all from Navy families and proud of it. I think the movie house held maybe fifty people at any given time. The pool was not much bigger, but it got a lot of use by everyone. I think the theater maybe charged 10 cents, but it might have been free. When they showed "Blackboard Jungle" there were about ten to fifteen people there with 90% of them being BHS kids. We could not believe what went on in the movie and the music. We loved it. And we were viewing it about six months after the States. When it was over three of us went to ask the old guy who ran the movie if he would show it again. The old guy (I think he may have been 22) said sure. So the old movie theater rocked till about 12 midnight. It was fun, but the next Ft. Kobbe Bulletin said the Far Fan theater movies were shown 1-ONE-time only.
- *Frank Mott*

Fort Amador/Fort Grant

Established in 1911, Fort Amador, located on the mainland across from the islands of Fort grant, was designated as the headquarters for the U.S. Army Caribbean in 1949. Fort Amador and Fort Grant Military Reservations were officially "set apart and assigned to all the uses and purposes of a Military, Reservation,"10 and their limits were defined, by Executive order #3130 on July 25, 1919. Although their jurisdiction ultimately fell under the control of the Secretary of War, both reservations were locally "subject to the civil jurisdiction of the Canal Zone authorities in conformity with the Panama Canal Act."

Forts Amador and Grant were assigned names by Secretary of War Henry L. Stimson in January of 1912, Fort Amador was named in honor of Dr. Manuel Amador Guerrero, the first-President of the Republic of Panama, at the suggestion of the United States Minister to Panama.

The two forts initially claimed only about 70 acres of land, but this expanded over the years to over 344. Fort Amador was the primary infantry and support area, and grew to include a rather prominent "tank farm" for fuel storage.

The Post Canal Construction Era (1912 - 1937)

New York Architect Austin W. Lord chose Itallanate Renaissance as the primary architectural style for the permanent buildings in the Canal Zone. Details of the Itallanate style include interior courtyards, large, often arched windows and verandas - features which capture breezes to cool the buildings' interiors - as well as heavily bracketed roofs. It was also a style popular in the United States in the early Twentieth Century.

 Memories

In 1972 we moved to a single family home, 0944 Amador Road. We were about five houses from the Amador gate. As kids we would ride our bikes to the Amador Golf Club and buy their hot dogs. They were so good because they had a machine that kept the bread warm. We would ride around by the gazebo and officers club, down to the causeway. My brother Paul would walk to the gym the GI's used and played basketball with them. My brother Paul William was born while we lived there.
- *Helen Hurst Loera*

The Bandstand at Fort Amador.

World War Two Construction Era (1938 - 1946)

World War II Era construction was in reaction to the anticipated increase in the number of troops required for Canal defense. Typically constructed with wood framing, these structures were intended to last only a few decades. With a few exceptions, emphasis was placed on function rather than aesthetics.

Casa Caribe, the six-unit Distinguished Visitors' Quarters, was constructed in 1939. Typical of the World War II Era, the two-story, raised structure is of wood frame construction. The Fort Amador Officers' Club was constructed in 1941 as a bowling alley.

With an increasing emphasis on sports, a baseball field was laid out on the Parade Ground at Fort Amador. The baseball field was named McCardell Field in January of 1957, in honor of Major Norman C. McCardell, U.S. Army Caribbean Special Services, who died on December 7, 1956, at Gorgas Hospital at the age of 39.

Contemporary Construction Era (1946 to 1970s)

The buildings during this time were designed by District architects and engineers of the Corps of Engineers. The design of these generic structures was intended to be international - that is, the buildings could be constructed at any military facility in the United States or in the world. Little emphasis was placed on environment or locale. The structure would protect the user from the cold climate of Alaska or the torrid heat of the Philippines - whichever were required. Purely functional, little emphasis was placed on aesthetics.

The Fort Amador Beach Club, located on the causeway, "was a favorite eating place for high school kids who wolfed their burgers ('the best hamburgers in town'), then went for a walk on the beach before returning to class."

This street scene exemplifies the typical planting schemes for residential areas at Fort Amador.

Painting of Amador Golf Course

Ft. Amador, named after Panama's first president, Manuel Amador Guerrero, was a United States Army base created during the construction days from excavated material taken from the Panama Canal. It is situated on Panama Bay overlooking the entrance to the Canal and is connected to the islands of Naos, Perico and Flamengo by the popular causeway that also was created from the canal excavation.

In 1936, a golf course was laid out and a clubhouse was constructed. Bordering the rocky coast of Panama Bay, holes # 3, #4 and #5 struck fear into golfers' hearts as they lined up their shots trying desperately to avoid the rocks and pounding surf from claiming their errant golf balls.

Bill Young (in foreground) preparing to play his second shot over the ravine on the feared hole #4 at Amador Golf Course. (Others, from L to R): Fred Sapp and Lou Engelke (on hole #6); Tom Pierce; Young; caddy, "Harry," and Ray Laverty (on green #4); John Bing and Clayton Murphy; Biff Clarke.

For many years, a painting depicting hole #2 hung in the Amador Golf Course bar and grille. In the early 80's, well-known American artist and General Counsel of the Panama Canal, John Haines, restored the painting to depict a scene that included the Panamanian skyline in the background with hole # 4 in the foreground and well-known local golfers on the course.

In 1989, during the US invasion of Panama, known as operation "Just Cause," the painting was removed temporarily for safekeeping by enterprising golf aficionados and reappeared later when normalcy was restored in Panama.

Upon the closing of the Amador Golf Club in 1998, the painting found its way to the Panama Canal Museum.

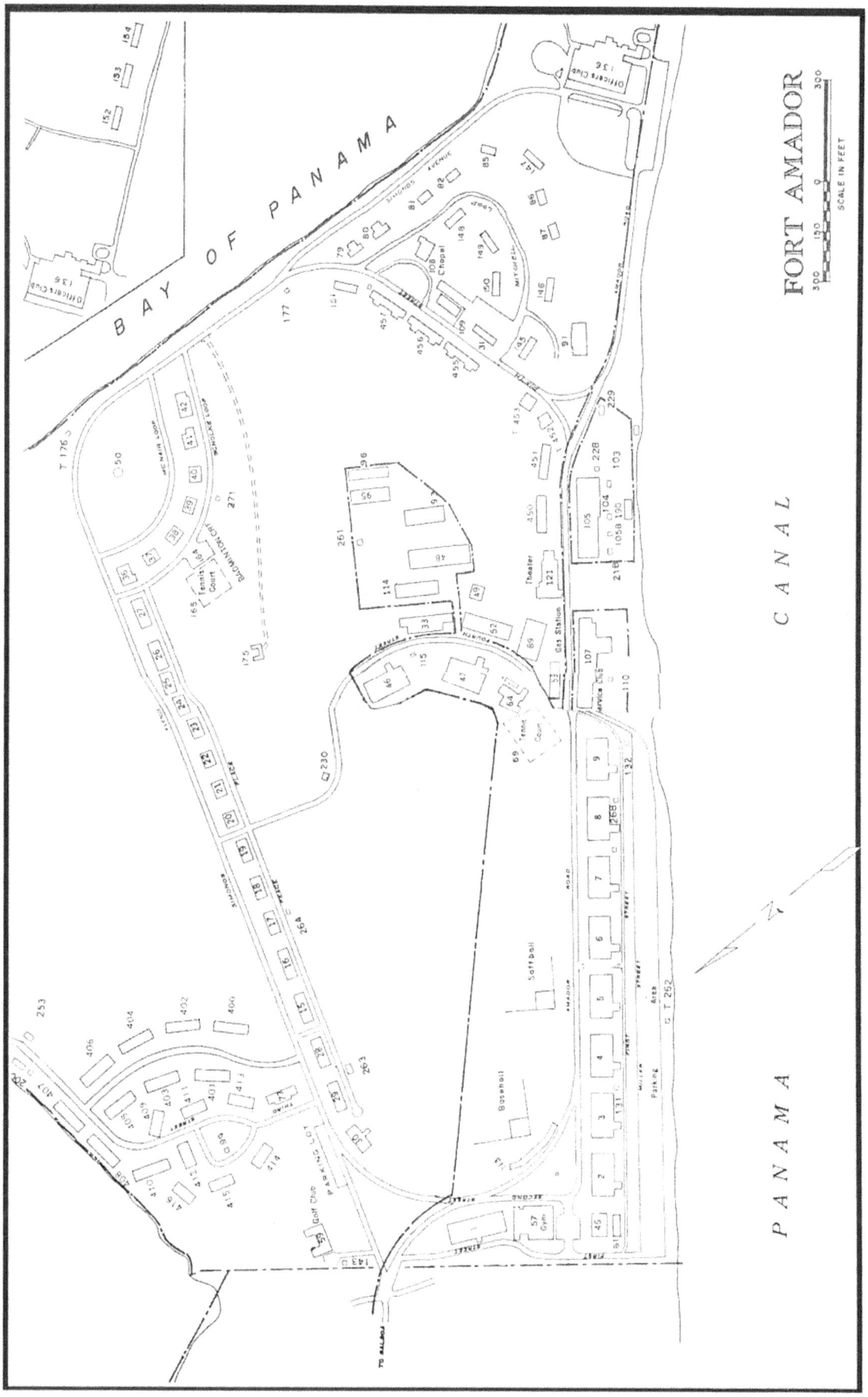

BAY OF PANAMA

FORT AMADOR

SCALE IN FEET

300   150   0   300

PANAMA

CANAL

PANAMA

154
153
152

136 Officers Club

135 Officers Club

85
82
81
80
73
47
96
48
108 Chapel
148
143
50
146
91
43
31
109
451
456
457
453
177
181
229
451
451
T451
450
103
228
96
95
9
97
114
49
251
261
105
218 104 190
105B
121
Theater
33
52
69
Gas Station
107
42
41
40
39
38
37
36
27
26
25
24
23
22
21
20
19
18
17
16
15
28
29
230
263
50
271
165 Tennis Court
175
115
46
47
54
59 Tennis Court
32
53
110
Service Club
9
8
7
6
5
4
3
2
132
12BB
131
45
57 Gym
253
406
404
402
400
407
409
403
401
413
414
411
412
410
415
414
55
143
WILLIAMS ROAD
AMADOR ROAD
FIRST STREET
SECOND STREET
FIRST STREET
Parking
Arch
Parking Lot
Golf Club
Softball
Baseball
McNAIR LOOP
GAILLARD LOOP
BLONDINGTON DR.
GRANT AVENUE
JAMES PLACE
PEARY PLACE
SIMMONS PLACE
26A
MITCHELL
LOOP
JANES PLACE
FOURTH AVENUE
T 176
T 453
T 252
N
TO RAILROAD
147

Fort Clayton

Sponsor

Ed English, U.S. Army, Retired
Tamara Martinez Gramlich

Aerial view of Soldier's Field Quadrangle with rear of Building 129 in the foreground.
From Guarding the Gates.

• Fire station, guest quarters, 2 elementary schools, 50-meter pool, 876 seat theater, 8-lane bowling alley with fast-food facility, automobile service garage, and a 10-lane gas station

Fort Clayton was one of the last U.S. military installations active in the Republic of Panama. In a departure ceremony on July 30, 1999, the Commanding General of U.S. Army South and Joint Task Force at Fort Clayton. Major General Phillip R. Kensinger Jr., said:

"All of you . . . before you leave, will probably want to walk or drive about the Fort a bit . . . to capture in your mind and heart a final picture - a well trimmed school yard where your children played – an empty stable where horses once lopped about – a set of quarters that you once called home. I'd also like to add another place, one which isn't even on

Fort Clayton, located on the east bank at the southern end of the Panama Canal, was built to defend the Miraflores Locks. Named in honor of Col. Bertram T. Clayton, Quartermaster of the Armed Forces in the Canal Zone from 1914-1917, served as headquarters for U.S. Army South (USARSO). Fort Clayton's assets include:
• 1392 family housing units and 1754 dormitories
• 410,000 square feet of warehouses
• 264,000 square-foot former hospital
• 150- and 375-person capacity community clubs with kitchens

The 876-seat Fort Clayton Theater.

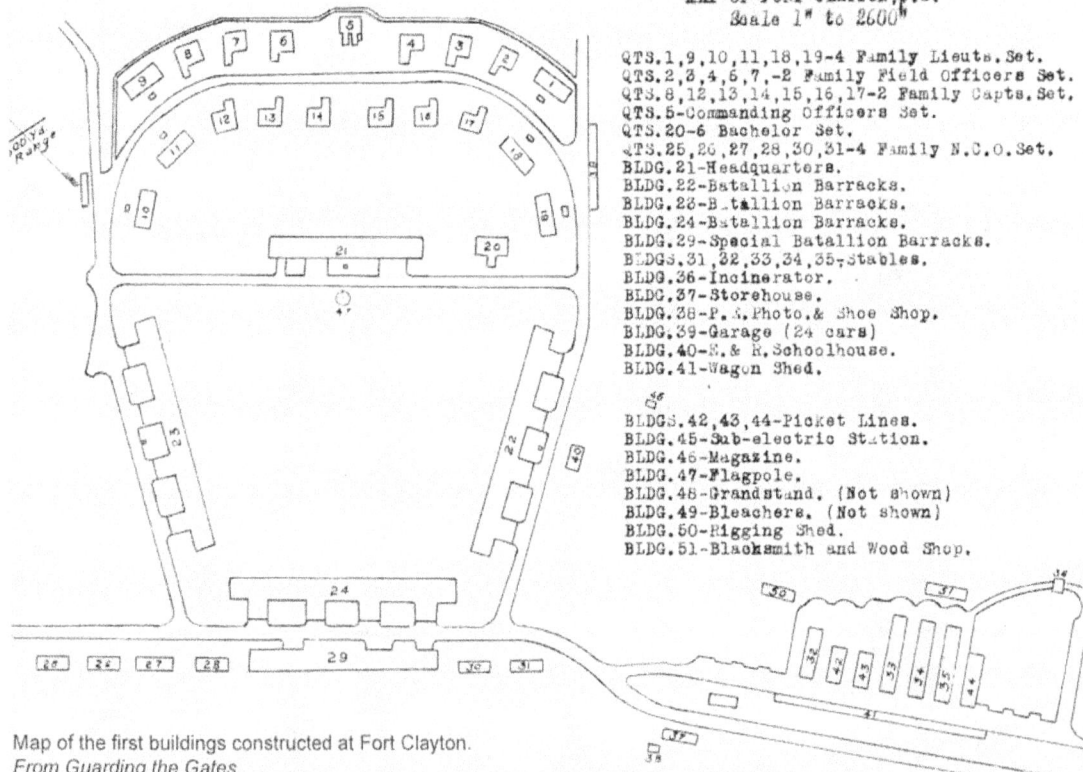

MAP OF FORT CLAYTON, C.Z.
Scale 1" to 2600"

QTS.1,9,10,11,18,19-4 Family Lieuts. Set.
QTS.2,3,4,6,7,-2 Family Field Officers Set.
QTS.8,12,13,14,15,16,17-2 Family Capts. Set.
QTS.5-Commanding Officers Set.
QTS.20-6 Bachelor Set.
QTS.25,26,27,28,30,31-4 Family N.C.O. Set.
BLDG.21-Headquarters.
BLDG.22-Batallion Barracks.
BLDG.23-Batallion Barracks.
BLDG.24-Batallion Barracks.
BLDG.29-Special Batallion Barracks.
BLDGS.31,32,33,34,35-Stables.
BLDG.36-Incinerator.
BLDG.37-Storehouse.
BLDG.38-P.X.Photo.& Shoe Shop.
BLDG.39-Garage (24 cars)
BLDG.40-E.& R.Schoolhouse.
BLDG.41-Wagon Shed.

BLDGS.42,43,44-Picket Lines.
BLDG.45-Sub-electric Station.
BLDG.46-Magazine.
BLDG.47-Flagpole.
BLDG.48-Grandstand. (Not shown)
BLDG.49-Bleachers. (Not shown)
BLDG.50-Rigging Shed.
BLDG.51-Blacksmith and Wood Shop.

Map of the first buildings constructed at Fort Clayton.
From Guarding the Gates.

Clayton, the Miraflores locks just across the way. From there you can see those ingenious locks that enable the canal to operate -- and you can see the Gaillard cut through the mountains that only the United States could make - and, finally, you can turn back and see Fort Clayton, its flag flying high and proud in the humid, tropical air of Panama. A canal, a country, a fort. The names are easy to remember. The sight is one that will live in your heart forever."

The American flag flying over Fort Clayton was lowered for the last time in a ceremony on November 30, 1999. For almost 80 years Fort Clayton had helped guarantee safe transit of the Panama Canal. And it had been home to many.

Fort Clayton was built on the site of the construction era's Miraflores Dump – a repository for excavated material adjacent to the Miraflores Locks. Beginning in 1915, there was a series of efforts to develop a plan for canal defense that included permanent military installations in the Canal Zone. Beginning in 1918, parcels of land were designated as U.S. military reservations through a series of presidential executive orders. In 1919 the Curundu Military Reservation was established; this included Fort Clayton, a new Army post at Miraflores Dump. This new post was named in honor of the late Col. Bertram T. Clayton who had served as Quarter Master of the Canal Zone military forces from 1914 to 1917.

The Panama Canal's Building Division designed (Office Engineers Section) and constructed the original installation at Fort Clayton. By the end of 1920 the first phase was done and Fort Clayton was declared ready for occupancy. The Infantry took possession on October 25, 1920. By 1922 construction of the installation was competed and "the soldiers of Fort Clayton assumed their role as guardians of the Miraflores and Pedro Miguel Locks." There was no significant new construction during the 1920s. However, Fort Clayton became a busy installation. Training for the protection of the canal locks was paramount. But Fort Clayton was also a place to live. By 1925 the enlisted men and officers at Fort Clayton had 2 tennis courts, a polo team, a playing field, and dayrooms with pool tables and reading material. Sports, horse shows and rodeos contributed to the competitive spirit and entertainment of the residents. And the 33rd Infantry Band provided additional entertainment.

Beginning in the 1930s, for over five decades Fort Clayton saw new construction. With WWII, Fort Clayton was expanded to accommodate wartime troop build up. "New construction took place on and around a high hill across the Cardenas River from all other post buildings and this left an empty area . . . for future expansion."

In the two decades after WWII, Fort Clayton became "a company town whose main business, as always, was to guard the gates between the two oceans." Construction programs provided the fort with all the amenities of an American town: a neighborhood school, a dispensary (located in the Fort Clayton Sector Hospital that had been opened in 1943 and inactivated

First wartime housing to be completed: the 300 Area NCO quarters. *From Guarding the Gates.*

as a hospital in 1954), a commissary, a PX, social clubs, a new 50-meter Olympic-size swimming pool complex, a new gymnasium, several softball fields and grandstands, a bowling alley, a golf course and club house, two theatres, a chapel, and family housing that included sets of officers' quarters and sets of enlisted men's quarters. When the Panama Canal Treaties were signed in 1977 (the "1977 Treaties"), Fort Clayton was in its almost-final form.

After WWII, in an attempt to retain officers and enlisted men on a career basis, the military had embarked on a program of providing service personnel with attractive family homes. As a consequence, post-WWII additions to Fort Clayton built an environment that resembled suburbia with curvilinear streets, housing clustered off main roads, cul-de-sacs and landscaping that provided shading, decoration and privacy. Even conveniently located schools for young children were ultimately included.

Until the 1962-63 school year, all the children of Fort Clayton attended Canal Zone schools off-installation. In the 1962-63 school year, Fort Clayton's need for a "local" elementary school was partially met when an old barracks was converted for that purpose. In 1967, a newly constructed Curundu Elementary School opened on a tract of land near the back gate of Fort Clayton. Curundu Elementary was managed by the Panama Canal Company (PCC) until transfer of the school to the Department of Defense Dependents' School (DoDDS) upon implementation of the 1977 Treaties. When the school property was transferred to Panama in 1999, its total value was assessed at $1,569,091.

As to other quality-of-life improvements, over time, electrical service was

upgraded from 25-cycle to 60-cycle. This allowed for window-mounted air conditioners and other appliances (such as TVs) commonly in use in the United States

Indeed, the transmission station for American radio and television programs was located in Fort Clayton. A permanent radio-transmitting station was built at Fort Clayton in December 1941. During the war years and thereafter, Zonians and military alike enjoyed baseball games, radio programs, music and news from the United States broadcast over the Armed Forces Radio Station (AFRS). In the 1950's the Southern Command Network (SCN) radio station on Fort Clayton was outfitted to become the first television station in Panama. Broadcasts from Fort Clayton continued until its transfer to Panama in 1999.

In 1978 the U.S. Senate ratified two treaties with Panama. The first, called the "Panama Canal Treaty," abolished the Canal Zone and turned the territory over to the Republic of Panama, the U.S. retaining the authority to operate and defend the canal until December 31, 1999. The second treaty provided for the permanent neutrality of the canal and granted the U.S. the permanent right to defend that neutrality.

On October 1, 1979, the treaties were put into effect. The Panama Canal Company and the Canal Zone Government were dissolved, and the Panama Canal Commission was established to run the Panama Canal for the next 20 years. Fort Clayton (including Corozal) was re-designated as one of the "Defense Sites" to remain under the control of U.S. forces until December 31, 1999. It became the only Army Defense Site on the east bank of the Pacific side of the canal. As a consequence, many activities and units were relocated to

Building number 208, NCO single quarters at Fort Clayton. *From Guarding the Gates.*

Fort Clayton, and it was renamed Headquarters 193rd Infantry Brigade (Panama). In 1986, a reorganization made Fort Clayton headquarters for a reactivated United States Army South (USARSO) new major command.

During this 20-year period the emphasis on defense of the canal remained primary for Fort Clayton. However, other missions emerged. The military was used to help fight the war on drugs, support assistance programs in Latin America, provide disaster relief and search-and-rescue, and provide support for inter-American training exercises. Fort Clayton played a key role in "Operation Just Cause" in late 1989-early 1990.

In December 1989, tensions between the United States and Panamanian dictator Manuel Noriega came to a head. Following Noriega's annulment of May 1989 election results that would have him replaced as President of Panama, the United States denounced him and called for his overthrow. On December 15, 1989, Noriega declared that Panama was at war with the United States. On December 17, 1989 President George H. Bush ordered "Operation Just Cause." This operation was successful, ending with the detention of Noriega and the installation of the new Panamanian president.

Fort Clayton was central to this success. Headquarters for the entire operation was at Fort Clayton. Moreover, Fort Clayton actively protected the locks with combat troops, suspending transit operations for only 30 hours. Finally, Fort Clayton hosted the administration of the oaths of office to the president-elect and vice presidents of Panama.

By the end of 1999, Fort Clayton had performed successfully for almost 80 years its mission of defending the canal. The canal had never been damaged by sabotage or military attack. The final value of the facilities at Fort Clayton at turn-over in 1999 was $119,943,000.

Top, Clayton Towers condos. Above, World War II Fort Clayton Hospital. *Courtesy of Ed English.*

Left, Post Headquarters, Fort Clayton. Courtesy of Isabel Wood Egan. Right, Aerial view where its 6,056 acres included 1392 housing units and 1754-person dormitories as well as a hospital, clubhouses, a fire station, 2 elementary schools, a 50-meter pool, theater and 8-lane bowling alley.

U.S. Naval Air Station at Coco Solo

Established in 1918, Coco Solo was a United States Navy submarine base on the Atlantic side of the Panama Canal Zone. In the summer of 1941, Coco Solo Hospital was built after the area was transferred from the civil part of the Panama Canal Zone when Franklin Roosevelt signed Executive Order 8981 on December 17, 1941. During World War II, Coco Solo served as a Naval Aviation Facility housing a squadron of P-38 Lightning aircraft.

During World War II a 200-bed hospital was built on the base which was later transferred to the Panama Canal Company in 1954. By the 1960s no U.S. Navy vessels remained, only some support staff and housing.

Fort Davis

Sponsors

Ed English, U.S. Army, Retired
Charles A. Thomas
J. E. Dorn Thomas

Located on Gatun Lake near the Gatun locks on the Atlantic entrance of the Panama Canal, Fort Davis was known as "Camp Gatun" when it was stablished in 1919 as a military reservation for the defense of the Atlantic side of the Panama Canal.
Named in honor of Colonel William D. Davis, 361th Infantry, Ft. Davis' construction began in 1920. Provisions were made for quarters and barracks for one regiment of Infantry. Its 4,075 acres provided the army with a training area for jungle warfare courses and special forces training. and the entire complex included 427 buildings by 1995. Fort Davis was transferred to the Republic of Panama in September 1995. *From the Panama Canal Museum Collection, George A. Smathers Libraries, Univerisity of Florida.*

Army Housing at Fort Davis.
Courtesy of Ed English.

Fort Gulick

Sponsor

Ed English, U.S. Army, Retired

Established in 1941, Fort Gulick provided a central ammunition storage area for the Atlantic entrance to the Panama Canal, accommodations for large numbers of anti-aircraft personnel during World War II and a 1,000-bed hospital occupying 14 separate barracks.

In May 1962, the advance party from Company D, 7th Special Forces Group, Fort Bragg, NC, departed for Fort Gulick, Panama, in the Canal Zone, to establish the 8th Special Forces Group, the US Army's only full Special Action Force. In addition to line Special Forces companies, the SAF included a Military Intelligence detachment, a Medical detachment, a Military Police

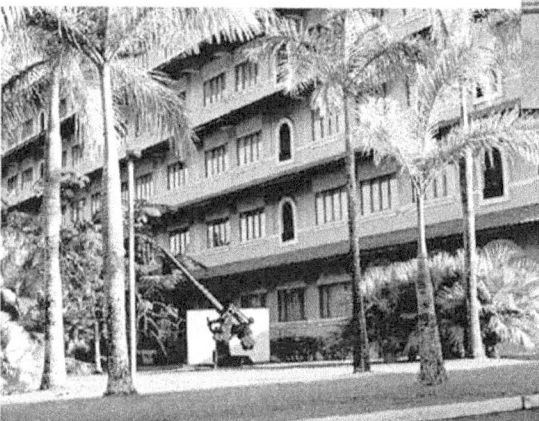

detachment, an Engineer detachment, a Security Agency detachment, and a Psychological Operations battalion. 8th Group was deactivated in 1972 and the unit reverted back as the 3rd Battalion, 7th Special Forces Group.

Above, Fort Gulick headquarters.

Left, School of the Americas, Melia Hotel. The School of the Americas also operated a campus at this base until 1984. That building is now a Hotel Melia. The fort was ceded to Panama in the early 1980's and became Ft. Espinar. After the 1989 invasion of Panama the fort was reactivated by the US as Fort Gulick/Espinar but was disestablished in September 1995. *Courtesy of Ed English.*

Left, Fort Gulick housing. Right, Fort Gulick Theater. *Courtesy of Ed English.*

Fort Gulick

The top floor of the USARSA building had always been rumored to be haunted. Originally built as a hospital, the story was that a nurse had died during WWII and roamed the halls of the fifth floor. On one particular evening, Greg Wolf, Wayne Forrest, Tank Davis, and I decided that it was time to confirm or dispel the legend.

It was the spring of 1980 and beer-induced courage was running high. We snuck past the guard at the desk who had to be in a coma not to hear us; we took the elevator to the fourth floor. It didn't go to the top as the fifth floor was closed—undoubtedly, we believed, due to supernatural activity.

We climbed the stairs to the top with little regard for how much noise we were making, which did not amuse Wolf because his Dad was in charge at USARSA. We tried to convince him that his Dad wouldn't mind and that the guy at the desk had obviously slept through Pearl Harbor so it wouldn't be a problem. He, however, didn't believe us.

We burst through the door into the fifth floor hallway and immediately got a little spooked. The lighting was very dim, noises and creaks were everywhere which made us carry on even louder because as everyone knows, beer dulls your hearing. The noise was bothering Wolf who was still caught up in the whole "my Dad will kill me" thing, so he began to make us be quiet. Did I mention that Wolf was possibly the largest person on the Atlantic Side.

At this point, the courage turned to fear-induced hysterical laughter and Wolf "decided" it was time to leave. We rumbled down the stairs to the elevator, piled in and suffered the close quarters with Wolf until the doors opened at the bottom. The small stampede through the lobby finally stirred Rip Van Winkle, but by the time he vacated his chair, we were out of the parking lot.

We never solved the mystery, but at least we weren't in custody!
- Kurt McQuillen

Fort Gulick Pool. *Courtesy of Ed English.*

Fort Kobbe / Howard AFB

Implementation of the 1977 Panama Canal Treaty: An Overview, p. 16.

A 1,804 acre fort established in 1928, Fort Kobbe is located strategically on the western bank of the Atlantic entrance to the Panama Canal. It was named in honor of Major General William A. Kobbe, a distinguished artilleryman known to have

written much of the Army's artillery doctrine. The site is characterized by rocky coastline and two beaches with fresh water sources, and defending the Bay of Panama was the site's original mission. Prior to WWII, little development beyond installation of big guns and an unpaved airstrip had occurred. The large and outdated guns were removed and $14 million was spent to develop a more complete artillary post during WWII. Eventually an improved airstrip gained its own status as a separate installation and was named Howard Air Force Base for aviation service pioneer Charles Harold Howard.

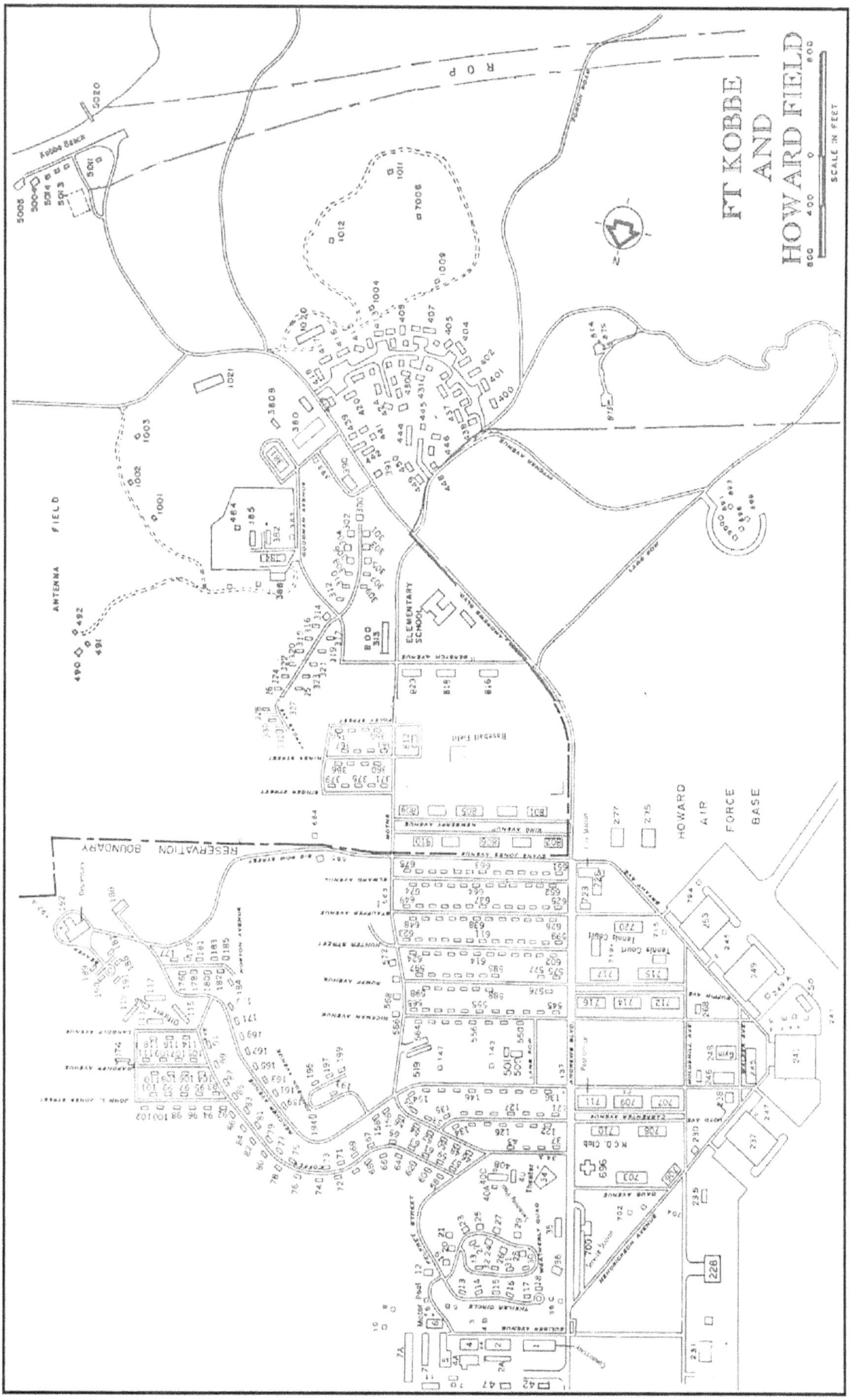
FT KOBBE
AND
HOWARD FIELD
800 400 0 800
SCALE IN FEET
ANTENNA FIELD
RESERVATION BOUNDARY
HOWARD
AIR
FORCE
BASE
ELEMENTARY SCHOOL
BOQ 313
Baseball Field
Theater
N.C.O. Club
Tennis Court
Kobbe Beach
R O P
Motor Pool

Memories

I remember when I lived in Howard for about five years. We used to do just a little bit of sliding down some hills. Our favorite hill was Suicide Hill; it was back behind the New Housing area in Howard Air Force Base. We mainly went on the hill when rainy season came. The gang and I used to slide down the hill with the garbage tops; sometimes we slid down by the back of our pants. After we finished we went into the jungles and played Army. We always split everybody up for different teams. Then we tried to find the opposite team and kill them before they killed us. I always took my dog with us so we could find the other people.

When it started to get dark, we went home and ate. Then we played Kick the Can or Ring Door Bells for a little entertainment around the block. About nine o'clock we had to go in.
- *David Peck in 1978 for a folklore class taught by Mary Knapp at Balboa High School.*

Fort Sherman

Sponsor

Ed English, U.S. Army, Retired

Fort Sherman included 23,100 acres (93 km2) of land, about half of which was covered by jungle. Located on the Atlantic side of the canal. The first U.S. Army troops including infantry, cavalry, engineer, signal, and field artillery units made up what was known as the Mobile Force, began arriving in October 1911. Giant artillery guns were mounted in massive concrete emplacements. In 1941, Fort Sherman was the site of the US's first operationally deployed early warning radar.

The base facilities included 67 family housing units, barracks for 300, a 2,775-foot airstrip, and various recreational areas including a pool, gym, beach. It also had a 200-seat theater and a clubhouse with a kitchen.

Abandoned barracks at Fort Sherman in 2008.

A battery overrun by the jungle.

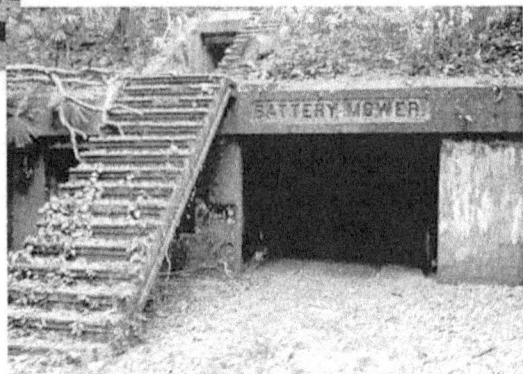

Jungle Operations Training Center was created in 1951 to train both US and allied Central American forces in jungle warfare, with an enrollment of about 9,000 a year. The JOTC also taught a 10-day Air Crew Survival Course, open to all branches of service, and a four-week Engineer Jungle Warfare Course. The dock at Fort Sherman is now the Shelter Bay Marina, and much of the base is now reclaimed by the rain forest.

In 2008 Fort Sherman was used in the filming of the James Bond film *Quantum of Solace* .

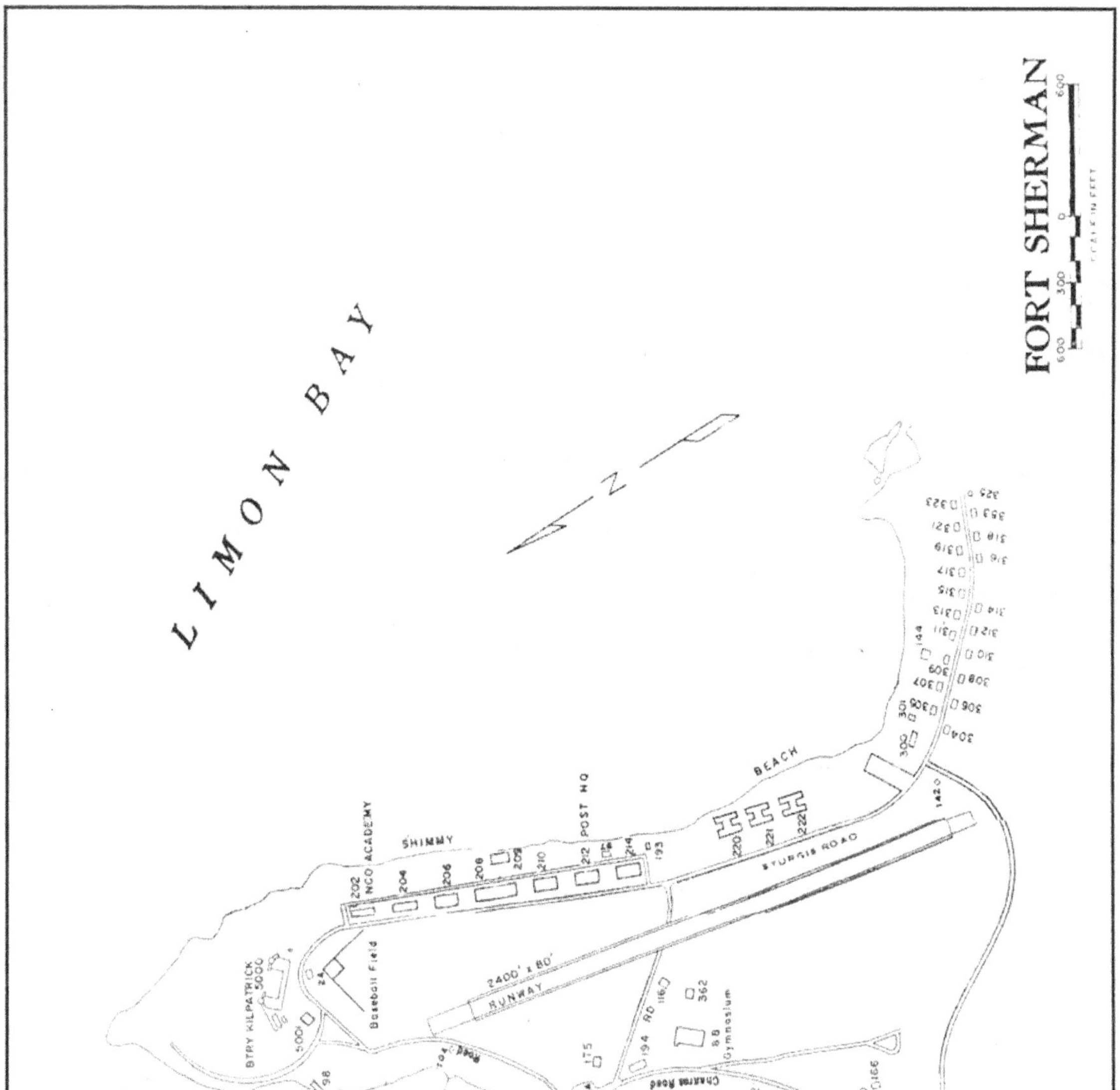

FORT SHERMAN
600 300 0 600
SCALE IN FEET
LIMON BAY
CARIBBEAN SEA
FORT SHERMAN LAGOON
SHIMMY BEACH
NCO ACADEMY
POST HQ
STURGIS ROAD
2400' x 80' RUNWAY
Baseball Field
BTRY KILPATRICK
Gymnasium
Chagres Road
MARTIN RD
CHAPEL
BUTNER ROAD
TORO PT ROAD
LIGHT HOUSE ROAD
KENNEDY LOOP
KESSLER PLACE
RIGBY PLACE
RIGGS PLACE

France Field

Sponsors

Ray & Justine Bunnell Family
Joseph H. White, Jr., & Family

France Field: Established 1917

Implementation of the 1977 Panama Canal Treaty: An Overview, p. 12

France Field was the first real air base in the Canal Zone. It encompassed 634,68 acres on Manzillo Bay near forts Randolph and De Lesseps. Pan-American Ariways flew bi-monthly flights beween here and Miami beginning in 1928, and U.S. Airmail service to the Canal Zone began in 1929.

France Field Quarters Returned To Canal

The Panama Canal Review, July 4, 1958, Page 13

Gold Hill, not the great rocky crag on the east side of Gaillard Cut, but one of the most attractive residential sites on the Atlantic side of the Canal Zone. Gold Hill will be reopened for occupancy by Company-Government employees within the next few months.

Gold Hill lies east of Randolf Road in what was once the France Field military reservation and contains a set of fine masonry houses. The quarters have passed back and forth between the military and the Panama Canal for years.

This week Gold Hill again passed into the hands of the Canal organization, this time permanently as this area will now be excluded from the military reservation. Work of rehabilitating the quarters will be started at once; the first of them should be ready for assignment about September 1.

The 29 apartments at Gold Hill, 15 of them single houses and the remainder in seven duplexes. This will provide sufficient housing for the Atlantic side of the Zone to permit all American families and most bachelors to be moved from 12-family buildings, and will round out the housing replacement program for the Atlantic side. The Gold Hill quarters were inspected by a committee from the Board of Directors during their visit here last month.

Built between 1936 and 1942, the quarters were first occupied by the top "brass" stationed at France Field. In August 1950, the houses were transferred to the Canal organization on a temporary status and were occupied by Canal families—including the Tatelmans, the Tiptons, and many other well-known Atlantic siders—until the Armed Services again required the quarters in June, 1953.

All of the apartments are in two-story masonry buildings; some have a ground floor, or "basement"—Canal Zone style. Some of the single houses have three bedrooms; others have four bedrooms; all have two upstairs baths. The duplexes are all three bedroom apartments.

All of the quarters will require considerable rehabilitation. Vinyl tile flooring will be installed in the kitchen and pantry. All will be given complete exterior painting.

Except for two of the single houses, all of the apartments will be assigned in accordance with the normal assignment procedure. The two exceptions are permanent, official assignments to the Chief of the Industrial Division and the Cristobal Magistrate. As the quarters are ready for occupancy, they will be advertised on the regular quarters vacancy bulletin. The "two year rule" will not apply to the Gold Hill quarters.

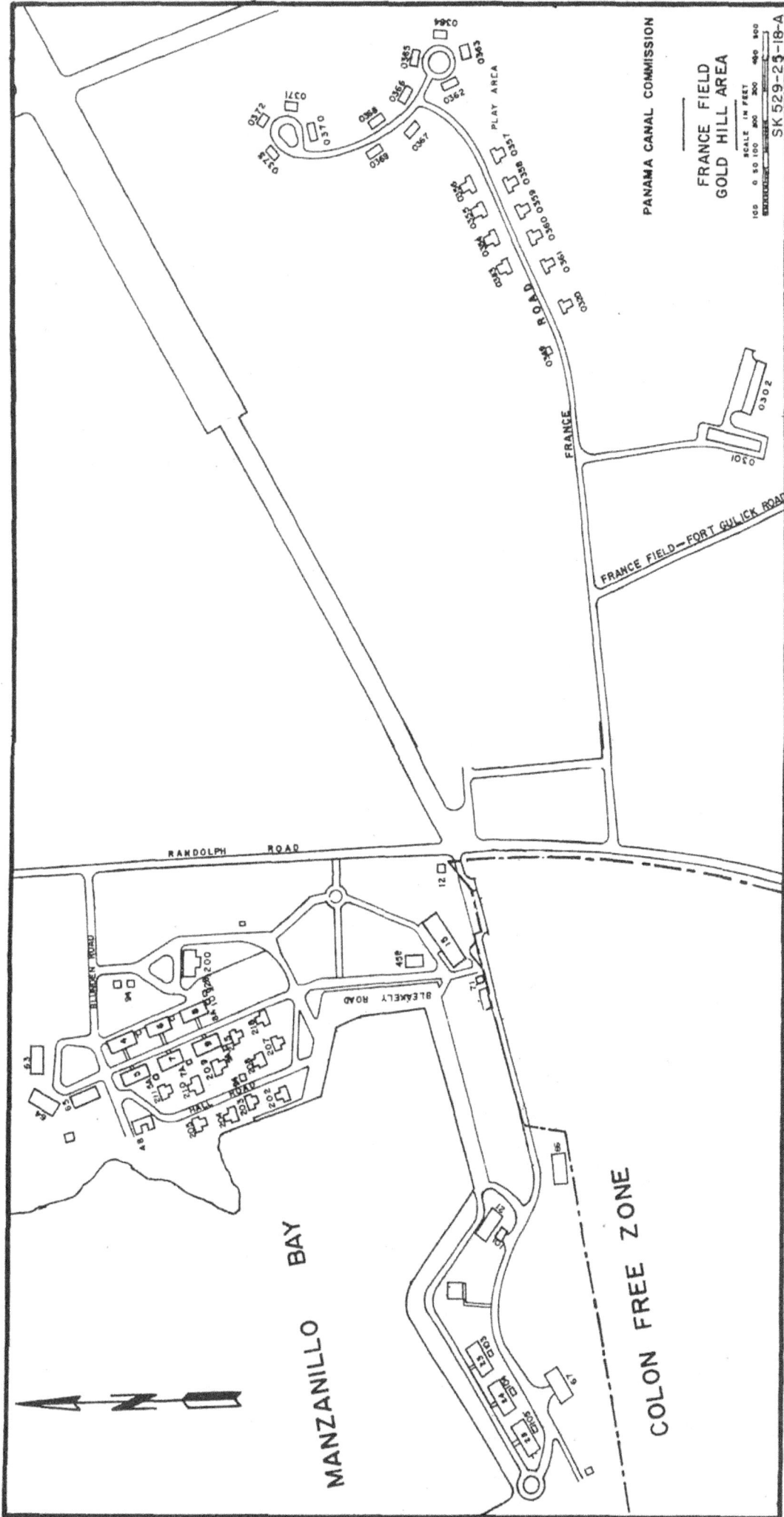

PANAMA CANAL COMMISSION

FRANCE FIELD
GOLD HILL AREA

SCALE IN FEET
100 0 50 100 200 300 400 500

SK 529-25-18-A

MANZANILLO BAY

COLON FREE ZONE

RANDOLPH ROAD

FRANCE FIELD—FORT GULICK ROAD

157

Quarry Heights

Sponsor

Col. James W. Wilson, USMC, Ret.

Quarry Heights had its beginning in February, 1909, when the Isthmian Canal Commission began quarrying rock for construction of the Miraflores and Pedro Miquel locks. Locals soon dubbed the area Quarry Heights. The Commission closed the quarry after five years, having removed more than 3.2 million cubic yards of rock. Canal officials transferred the quarry shelves to the U.S. Army in 1914 for use as a Canal Zone Army command post.

Quarry Heights was the nerve center for U.S. Military forces in Panama since 1919 and for all U.S. Forces deployed in the Southern Theater since 1947. Transferred to the Army by the Panama Canal Commission in 1914, Quarry Heights served as a troop encampment, military police post, and senior officer housing area for six years. During this time, the Army upgraded the post to make it suitable for use as the headquarters of the ranking military commander in Panama. Subsequently, Quarry Heights was the headquarters location for the Commander, U.S. Army Panama Canal Department (1920-1941); Commander, U.S. Army Caribbean Defense Command (1941-1947); Commander in Chief, U.S. Caribbean Command (1947-1963); and Commander in Chief, U.S. Southern Command (1963-1998). Quarry Heights' role as a military command center ended in 1998, when proprietorship of the reservation was passed to the Republic of Panama.

Quarry Heights Military Reservation is a picturesque historic district composed of buildings which date back to the Canal Construction Era. These buildings, primarily family housing units, reflect an architectural style unique to the Panama Canal area -- a style which incorporated both French and American design elements for living in a hot and humid tropical climate.

Quarry Heights is situated on two man-made terraces created by quarrying on the slope of Ancon Hill. The Hill, which rises 654 feet from the bay at its foot, has been a key geographical reference point in Panama for nearly 400 years. This is due to its proximity to Panama City and its commanding 360-degree view encompassing the Pacific Ocean approach to the Isthmus of Panama and the Panamanian hinterland seven to ten miles to the west, east, and north.

Above Left, Montague Hall, Command Headquarters building for the Caribbean Command known as USARCARB. Right, back side of Montague Hall with the Parade Field. Below left, Army barracks, which faced onto 4th of July Avenue. Below right, little building on the left next to the palm tree was the Post Theater. *Courtesy of Chuck Heller*

Rodman Naval Station

Sponsor

Charles W. "Chuck" Hummer

Implementation of the 1977 Panama Canal Treaty: An Overview, p. 17.

Established in 1932. Located on the western bank of the Panama Canal, Rodman Naval Station was named in honor of Captain Hugh Rodman, marine Superintendent and Superintendent of Transportation of the Isthmian Canal Commisson. Rodman Naval Station's valuable facilities include: deep draft port with 4,000 feet of pier space, a boating and recreation marina, retail stores, fire station, 50-meter pool, chapel, office building with post office, and a heavy industrial area. It also had 86 family housing units and 182-person dormitories.

Rodman, CZ. Family Quarters for CPO

Below, Military housing.
Courtesy of Bill Fall

Below, Swimming pool at Rodman.
Courtesy of Bill Fall

Bowling Alley at Rodman.
Courtesy of Bill Fall

Bibliography

American Women on the Panama Canal, edited by Mrs. Ernest Ulrich von Munchow, 1916.

The Canal Record, 1907-1908, Isthmian Canal Commission.

The Canal Review, Panama Canal Company and Panama Canal Commission,1950-1981.

Cocoli Verses: Cocoli, Canal Zone, Carl N Berg.

Guarding the Gates: The Story of Fort Clayton-Its Setting, Its Architecture, and Its Role in the History of the Panama Canal, Susan I. Enscore, Suzanne P. Johnson, Julie L. Webster and Gordon L. Cohen, US Army Corps of Engineers, 2000.

A History of Fort Amador and Fort Grant, Legacy Resource Management Program Demonstration Project, prepared for United States Army South (USARSO) through the Directorate of Engineering and Housing, United States Army Garrison-Panama, by Graves-Klein, Architects, Engineers of Pensacola, Florida.

Implementation of the 1977 Panama Canal Treaty: An Overview.

Isthmian Folklore, Collected by the Students of Mary Knapp's Folklore Classes at Balboa High School, 1978.

The Panama Canal Spillway, 1996.

Recreation in Panama Canal Company/Government Communities, prepared by C. H. Raybourn and Debbie Henson, Canal Zone Government, Civil Affairs Bureau, 1973.

Schooling in the Panama Canal Zone, 1904 -1979, Phi Delta Kappa, Panama Canal Area, 1980.

Schooling in the Panama Canal Zone, 1904 -1989, Phi Delta Kappa, Panama Canal Area, 1989.

Townsites & Area Maps, Canal Zone, Panama and Colon.

Title Page Credits

Cover	One year old Norma in Gatun, courtesy of Norma Stillwell Martin Balboa, courtesy of Vicki Hutchison Boukalis.
Page 5	Construction Era Towns. *From the Hallen Collection.*
Page 27	Balboa. *From The Panama Canal Review,* March 7, 1952.
Page 45	Cocoli. *Courtesy of Carl Berg.*
Page 51	Coco Solito. *Courtesy of Mary Elizabeth (Beth) Bialkowski Lozano.*
Page 53	Coco Solo - *From The Panama Canal Review.*
Page 61	Cristobal. New Cristobal, Old Cristobal. *Drawing of Cristobal High School, Courtesy of Fred Raines.*
Page 85	Gatun. *From the Hallen Collection.*
Page 95	La Boca. *From the Hallen Collection.*
Page 101	Los Rios. *From the Hallen Collection.*
Page 107	Margarita. *From the Hallen Collection.*
Page 115	Pariaso. *From the Hallen Collection.*
Page 121	Pedro Miguel. *From the Hallen Collection.*
Page 127	Rainbow City. *From the Panama Canal Review.*
Page 133	Red Tank. *From the Hallen Collection.*
Page 137	Military Townsites. *Panama Canal Department, Army Enlistment Poster,* from *Guarding the Gates.*

www.ingramcontent.com/pod-product-compliance
Lightning Source LLC
Chambersburg PA
CBHW081622280326
41928CB00056B/2886